NOMENCLATURE OF INORGANIC CHEMISTRY II

IUPAC PERIODIC TABLE OF THE ELEMENTS

1	2	3	4	5	6	7	8	9	10	11	12	13	14	15	16	17	18	n
1 H																	2 He	1
3 Li	4 Be											5 B	6 C	7 N	8 O	9 F	10 Ne	2
11 Na	12 Mg											13 Al	14 Si	15 P	16 S	17 Cl	18 Ar	3
19 K	20 Ca	21 Sc	22 Ti	23 V	24 Cr	25 Mn	26 Fe	27 Co	28 Ni	29 Cu	30 Zn	31 Ga	32 Ge	33 As	34 Se	35 Br	36 Kr	4
37 Rb	38 Sr	39 Y	40 Zr	41 Nb	42 Mo	43 Tc	44 Ru	45 Rh	46 Pd	47 Ag	48 Cd	49 In	50 Sn	51 Sb	52 Te	53 I	54 Xe	5
55 Cs	56 Ba	*57 La	72 Hf	73 Ta	74 W	75 Re	76 Os	77 Ir	78 Pt	79 Au	80 Hg	81 Tl	82 Pb	83 Bi	84 Po	85 At	86 Rn	6
87 Fr	88 Ra	‡89 Ac	104 Rf	105 Db	106 Sg	107 Bh	108 Hs	109 Mt	110 Uun	111 Uuu	112 Uub	113 Uut	114 Uuq	115 Uup	116 Uuh	117 Uus	118 Uuo	7

*57 La	58 Ce	59 Pr	60 Nd	61 Pm	62 Sm	63 Eu	64 Gd	65 Tb	66 Dy	67 Ho	68 Er	69 Tm	70 Yb	71 Lu	6
‡89 Ac	90 Th	91 Pa	92 U	93 Np	94 Pu	95 Am	96 Cm	97 Bk	98 Cf	99 Es	100 Fm	101 Md	102 No	103 Lr	7

International Union of Pure and Applied Chemistry

Nomenclature of Inorganic Chemistry II
RECOMMENDATIONS 2000

Issued by the Commission on the
Nomenclature of Inorganic Chemistry
and edited by J.A. McCleverty and N.G. Connelly

ROYAL SOCIETY OF CHEMISTRY

ISBN 0-85404-487-6

A catalogue record for this book is available from the British Library.

© 2001 International Union of Pure and Applied Chemistry

All rights reserved.

Apart from any fair dealing for the purposes of research or private study, or criticism or review as permitted under the terms of the UK Copyright, Designs and Patents Act, 1988, this publication may not be reproduced, stored or transmitted, in any form or by any means, without the prior permission in writing of the International Union of Pure and Applied Chemistry, in the case of reprographic reproduction only in accordance with the terms of the licences issued by the Copyright Licensing Agency in the UK, or in accordance with the terms of the licences issued by the appropriate Reproduction Rights Organisation outside the UK. Enquiries concerning reproduction outside the terms stated here should be sent to the International Union of Pure and Applied Chemistry.

Published for the International Union of Pure and Applied Chemistry by the Royal Society of Chemistry,
Thomas Graham House, Science Park, Milton Road, Cambridge CB4 0WF, UK
Registered Charity Number 207890

For further information see the RSC web site at www.rsc.org and the IUPAC site at www.iupac.chemsoc.org

Typeset by RefineCatch Limited, Bungay, Suffolk
Printed by Bookcraft Ltd

Contents

QD
149
I57
2001
CHEM

Principal authors of this Edition   vii
Preface   viii

II-1		**POLYANIONS** 1
II-1.1		Introduction 1
II-1.2		Numbering of condensed polyanions 3
II-1.3		Polyanions with six central atoms 6
II-1.4		Polyanions with the Anderson structure 10
II-1.5		Polyanions with twelve central atoms 11
II-1.6		Polyanions with eighteen central atoms 20
II-1.7		Conclusion 22
II-2		**ISOTOPICALLY MODIFIED INORGANIC COMPOUNDS** 23
II-2.1		Introduction 24
II-2.2		Classification and symbolism 24
II-2.3		Isotopically substituted compounds 26
II-2.4		Isotopically labelled compounds 27
II-2.5		Locants and numbering of isotopically modified compounds 34
II-2.6		Summary of types of isotopically modified compounds 35
II-3		**METAL COMPLEXES OF TETRAPYRROLES** 36
II-3.1		Introduction 36
II-3.2		The construction of systematic names 37
II-3.3		Trivial names 42
II-3.4		Less common structural types 43
II-3.5		Trivial names of porphyrins, chlorins, chlorophylls, bilanes, fundamental rings and related species 46

CONTENTS

II-4		HYDRIDES OF NITROGEN AND DERIVED CATIONS, ANIONS AND LIGANDS 54
II-4.1	Introduction 54	
II-4.2	Parent hydrides 54	
II-4.3	Cations 55	
II-4.4	Anions 56	
II-4.5	Ligands 56	
II-4.6	Organic derivatives of the nitrogen hydride ligands 61	
II-5		INORGANIC CHAIN AND RING COMPOUNDS 62
II-5.1	Introduction 63	
II-5.2	Unbranched chain and monocyclic compounds 63	
II-5.3	Branched chain and polycyclic compounds 75	
II-5.4	Conclusion 94	
II-6		GRAPHITE INTERCALATION COMPOUNDS 95
II-6.1	General considerations 95	
II-6.2	Vocabulary 96	
II-6.3	Classification of graphite intercalation compounds 98	
II-6.4	Non-formula-based description of graphite intercalation compounds 99	
II-6.5	Formulation of graphite intercalation compounds 100	
II-6.6	Structural notation 101	
II-7		REGULAR SINGLE-STRAND AND QUASI SINGLE-STRAND INORGANIC AND COORDINATION POLYMERS 104
II-7.1	Introduction 105	
II-7.2	Fundamental principles 106	
II-7.3	Recommendations 110	

INDEX 126

Principal Authors of this Edition

Chapter II-1
Y. Jeannin, M. Fournier, *France*

Chapter II-2
W.C. Fernelius, *USA*
T.D. Coyle, *USA*
W.H. Powell, *USA*
P. Fodor-Csányi, *Hungary*

Chapter II-3
R.S. Laitinen, *Finland*
J.E. Merritt, *USA*
K.L. Loening, *USA*

Chapter II-4
J. Chatt, *UK*
R.S. Laitinen, *Finland*

Chapter II-5
R.S. Laitinen, *Finland*
E. Fluck, *Germany*

Chapter II-6
H.P. Boehm, *Germany*
R. Setton, *France*
E. Stumpp, *Germany*

Chapter II-7
L.G. Donaruma, *USA*
B.P. Block, *USA*
K.L. Loening, *USA*
N. Platé, *Russia*
T. Tsuruta, *Japan*
K.Ch. Buschbeck, *Germany*
W.H. Powell, *USA*
J. Reedijk, *Netherlands*

Preface

This book is intended to be an adjunct to the first volume which appeared in 1990 and which contains aims, grammar, basic tenets and main classes of nomenclature. For these basic aspects the reader is referred to the primary volume. The current volume is intended to deal with the nomenclature of systems which have specific difficulties with condensed phases such as solids or polymers and where various local nomenclatures have grown up to confuse the practitioners or those entering the fields of research. It addresses such diverse chemistry as polyanions, isotopic modification, tetrapyrroles, nitrogen hydrides, inorganic ring, chain, polymer, and graphite intercalation compounds. It is aimed at bringing order into the nomenclature of these specialized systems based on the more fundamental nomenclature described in the primary volume and in the organic nomenclature publications. The volume is timely in relation to the growth of inorganic polymers especially and it is hoped that the practitioners will accept the orderly recommendations proposed and use it to simplify the literature especially for the students who now begin to study such systems.

The various chapters have been subject to extensive review not only by Commission members but also by practising chemists in the various areas over several years. The evolved chapters should therefore represent a consensus view of appropriate nomenclature for now and some years to come. The Commission is grateful to the external reviewers for their careful and critical comments and support to bring order into the nomenclature of such complex systems. Also, since the topics are largely unrelated and each has an Introduction or Preamble, no overall introduction is required and the reader is referred to the first volume for such general comments on nomenclature.

1st July, 2000

ALAN SARGESON
HERBERT D. KAESZ
Chairmen, IUPAC Commission of the Nomenclature of Inorganic Chemistry

Nomenclature of Inorganic Chemistry II
RECOMMENDATIONS 2000

II-1 Polyanions

CONTENTS

II-1.1 Introduction
II-1.2 Numbering of condensed polyanions
   II-1.2.1 Choice of reference axis
   II-1.2.2 Choice of preferred terminal skeletal plane
   II-1.2.3 Choice of reference symmetry plane
   II-1.2.4 Numbering central atoms
   II-1.2.5 Octahedron vertex designation
II-1.3 Polyanions with six central atoms
   II-1.3.1 Homopolyanions (isopolyanions)
   II-1.3.2 Heterocentre polyanions
      II-1.3.2.1 Mono- or polysubstitution
      II-1.3.2.2 Reduced heterocentre polyanions
   II-1.3.3 Heteroligand polyanions
      II-1.3.3.1 Single substitution
      II-1.3.3.2 Several substitutions
   II-1.3.4 Names of more complicated species
II-1.4 Polyanions with the Anderson structure
   II-1.4.1 Polyanions with seven central atoms
   II-1.4.2 Names of more complicated species
II-1.5 Polyanions with twelve central atoms
   II-1.5.1 Compounds with the Keggin structure and isomers
      II-1.5.1.1 Compounds containing only one kind of transition metal
      II-1.5.1.2 Compounds with several transition metals, *i.e.* substituted compounds
      II-1.5.1.3 Ligand substitution
      II-1.5.1.4 Reduced compounds
   II-1.5.2 Compounds in which central atoms are missing (defect structures)
      II-1.5.2.1 Compounds with one vacancy
      II-1.5.2.2 Compounds with three vacancies
II-1.6 Polyanions with eighteen central atoms
II-1.7 Conclusion

II-1.1 INTRODUCTION

A polyanion is formed by the condensation of several simple anions with the elimination of water. These negatively charged species have structures mainly made up of octahedra (polytungstates or polymolybdates), tetrahedra (polyphosphates), and sometimes octahedra and tetrahedra (polytungstates or polymolybdates). The octahedra and tetrahedra consist of a central atom surrounded by six or four atoms, respectively, which are referred to as ligands in this Chapter. The octahedra and tetrahedra share edges and vertices. The structure considered as an unsubstituted parent is the one which contains oxygen as ligands. Central atoms may be atoms of metals or, sometimes, non-metals. Some rare cases of 5-atom coordination and 7-atom coordination are known.

Either a central atom or a ligand can be replaced. Therefore, every atomic position must be numbered in order to be recognized and to distinguish isomers. In the nomenclature of coordination compounds, lower case letters have been suggested as locant designators for vertex designation. Central atoms have not commonly been given locant designators; however, number locants have been used for numbering metal atoms in homoatomic aggregates. In the first case, the position of a ligating atom of the ligands in the coordination polyhedra is given by a lower case letter. In the latter case, the ligand atom is indicated by a number which defines the central atom to which it is bound; if the ligand bridges several central atoms, several numbers are used. Thus, two locant systems presently coexist.

In the specific case of polyanions, difficulties arise because both central atoms and ligands can be replaced. The number of vertices of a condensed species is, in most instances, quite large: for example, $[SiW_{12}O_{40}]^{4-}$ has 40 vertices and is far from being the largest known polyanion. Obviously, the 26 letters of the alphabet are not sufficient if they are used for designating each vertex position. Since it is necessary to distinguish isomers, an unambiguous designation for central atoms, as well as for vertices, has to be devised. Moreover, the use of the numbers of two central atoms is not sufficient for designating bridging atoms because two or more bridges can occur between the same two central atoms.

The following numbering system is proposed:

(a) each central atom is given a number: 1, 2, 3, *etc.*,
(b) each polyhedron vertex is given a letter:
    octahedron – a, b, c, d, e, f
    tetrahedron – a, b, c, d.

A vertex is then designated by a number followed by a letter, the number referring to the central atom, *e.g.* 1a, 3d, *etc.* Thus, when two octahedra share a vertex, this vertex has two designations, one from the first octahedron, and one from the second octahedron, each octahedron surrounding its central atom. The designation with the lowest central atom number takes precedence. For example, if a vertex is 1d in the first octahedron and 4a in the second, it is designated by 1d.4a. Such a multiple designation might appear redundant. However, it may prove distinctly useful: for instance, in a discussion involving ligands located at vertices 1d and 4f, if 4a is an alternative for 1d, then 4a may be used instead of 1d to make it quite obvious that the two vertices, 4a and 4f, belong to the same octahedron. Moreover, this double designation makes it quite simple to name a common vertex: *e.g.* 1d.4a shows that vertex 1d is also 4a thus bridging central atoms 1 and 4 by their respective vertices d and a.

The numbering system used in this Chapter is consistent with the principles developed for boron cage compounds in Section I-11 of Note 1a and names are based on coordination nomenclature in the same book (Section I-10 of Note 1a), not on traditional oxoanion nomenclature, *e.g.* tetraoxophosphate(3–), not phosphate.

Note 1a. *Nomenclature of Inorganic Chemistry, Recommendations 1990*, Blackwell Scientific Publications, Oxford, 1990.

II-1.2 NUMBERING OF CONDENSED POLYANIONS

The numbering of a condensed structure is based on the unsubstituted parent structure for the polyanion. The central atoms of the octahedral units are numbered and the ligand positions are indicated by a secondary set of letter locants. Tetrahedral units are treated as bridging ligands.

Polyhedra constructed from octahedra contain symmetry axes of rotation and skeletal planes. Such planes are defined as those planes (or quasi-planes) containing several octahedral centres.

The following numbering recommendations are applied sequentially.

II-1.2.1 Choice of reference axis (see Figure II-1.1)

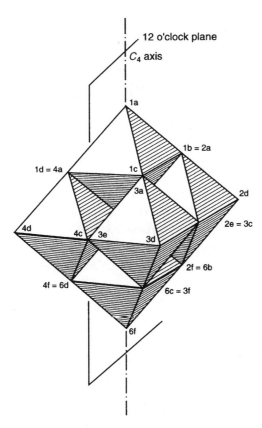

Figure II-1.1. Octahedron assembly and locant designators in the Lindqvist structure.

(a) The reference axis is the rotational axis of the polyanion structure of highest order; it is oriented vertically.

(b) Perpendicular to the reference axis, several skeletal planes may be encountered. A skeletal plane which lies farthest from the centroid of the polyanion is described as a terminal skeletal plane, others as internal skeletal planes.

(c) When there is more than one symmetry axis of highest order, the preferred axis is that which is perpendicular to the greatest number of skeletal planes.

(d) When the polyanion has no axis of rotational symmetry, the reference axis is then the axis perpendicular to the skeletal plane with the greatest number of octahedral centres.

II-1.2.2 Choice of preferred terminal skeletal plane

(a) The preferred terminal skeletal plane is that plane with the least number of central atoms. The reference axis is then oriented in such a way that the preferred terminal plane is uppermost.

(b) When both terminal planes contain the same number of central atoms, the preferred plane is that with the least condensed fusion of octahedra (*i.e.* when the number of bridges between cental atoms is the lowest; vertex sharing is less condensed than edge sharing which is less condensed than face sharing).

(c) See Section II.1.2.4 below.

(d) When a further choice is necessary, the preceding rules are applied considering the first internal skeletal planes, and so on.

II-1.2.3 Choice of reference symmetry plane

(a) The reference plane is defined as the symmetry plane which contains the reference axis and which also contains the lowest number of central atoms.

(b) When there is more than one reference symmetry plane which satisfies this requirement, then the preferred plane is that which contains the most atoms in common with the preferred terminal skeletal plane.

(c) The reference symmetry plane is divided by the reference axis in two halves which must be designated. A 6 o'clock – 12 o'clock line is defined by the intersection of the reference symmetry plane and a skeletal plane; thus it is perpendicular to the reference axis, and the 12 o'clock position is in the half of the reference plane which contains the largest number of central atoms; 6 o'clock designates the other half.

(d) See Sections II-1.2.4 and II-1.2.5 below.

(e) When a choice is left, the 12 o'clock position is chosen on a ligating atom.

II-1.2.4 Numbering central atoms

(a) Central atoms are numbered starting from the 12 o'clock position in the preferred terminal skeletal plane and turning clockwise (or anticlockwise). When a skeletal plane is fully numbered, the next skeletal plane located immediately below is numbered; the first atom to be numbered is that met when starting from the 12 o'clock position, turning clockwise or anti-clockwise, depending on the lowest locant requirement [see Sections II-1.2.4(b) and II-1.2.5].

(b) When a central atom or a ligand is substituted (see Section II-1.2.5), it does not lower the symmetry of the skeleton for the choice of the reference axis of symmetry and for the choice of the reference symmetry plane. However, locants are chosen in such a way that a central atom or a ligating atom appearing first in Table II-1 has the lowest number or the earliest letter. Nevertheless, the choice of the first skeletal plane as defined in Section II-1.2.2 takes precedence over the number of a substituting central atom. When the polyanion contains several central atoms of several different atomic species, the largest number of atoms of the species coming first in Table II-1 will be numbered before numbering an atom coming second in this Table, and so on.

POLYANIONS

Table II-1 Element seniority sequence.

He—Ne—Ar—Kr—Xe—Rn—
Li—Na—K—Rb—Cs—Fr—
Be—Mg—Ca—Sr—Ba—Ra—
Sc—Y—La→Lu
       Ac⇌Lr
Ti—Zr—Hf—Rf—
V—Nb—Ta—Db—
Cr—Mo—W—Sg—
Mn—Tc—Re—Bh—
Fe—Ru—Os—Hs—
Co—Rh—Ir—Mt—
Ni—Pd—Pt—
Cu—Ag—Au—
Zn—Cd—Hg—
B—Al—Ga—In—Tl—
C—Si—Ge—Sn—Pb—
H→N—P—As—Sb—Bi—
O—S—Se—Te—Po—
F—Cl—Br—I—At

II-1.2.5 Octahedron vertex designation

(a) In each octahedron letter locants are assigned to vertices as follows: define an axis passing through the central atom of the considered octahedron and parallel to the main reference axis. This defined axis is the local reference axis. These two axes make a new reference plane valid for this octahedron only. The vertices of the octahedron are in local skeletal planes perpendicular to these two axes. In each local skeletal plane, the local 6 o'clock – 12 o'clock line is the line intersecting the local reference axis and the main reference axis. The intersection with the main axis is the 12 o'clock position in the considered local skeletal plane.

Letters a, b, c, d, e, f are assigned starting from the upper local skeletal plane. The (possibly several) vertices in a given local skeletal plane are assigned letters turning clockwise around the local axis, starting from the local 12 o'clock position. If the local and the main axis coincide, then the local reference plane is the polyanion reference plane. The same set of rules is applied sequentially to give a letter locant designator to each vertex.

(b) If a choice exists in assigning letter locants to vertices, vertices are ordered according to the position in Table II-1 of the ligand atoms occupying them. An earlier position in this Table is assigned a letter earlier in the alphabet. In this connection, monoatomic ligands precede polyatomic ligands with the same ligating atom, *e.g.* oxygen atoms of OH or CH_3COO are considered to come immediately after oxygen and before any other element.

II-1.3 POLYANIONS WITH SIX CENTRAL ATOMS

The first representative example of a metal polyanion with six central atoms to have its structure determined was $K_8[Nb_6O_{19}]$ by Lindqvist. The idealized structure has O_h symmetry (Figure II-1.1). Several modifications of this structure are known:

(a) all central atoms are identical: such ions are commonly called isopolyanions; a better name is homopolyanions;
(b) one or more central atoms are substituted; these ions are commonly called mixed polyanions or heteropolyanions;
(c) some ligands are substituted.

Since substitution can occur either at a central atom or at a ligand site, these substituted ions are named as heterocentre polyanions or heteroligand polyanions, respectively.

II-1.3.1 Homopolyanions (isopolyanions)

In this structure, the metal atom has only oxygen ligands and there are six fused octahedra; it is sufficient to count the number of oxygens of each kind, *i.e.* of the same coordination.

Examples:
1. $[Nb_6O_{19}]^{8-}$
   μ_6-oxo-dodeca-μ-oxo-hexaoxohexaniobate(8–)

2. $[W_6O_{19}]^{2-}$
   μ_6-oxo-dodeca-μ-oxo-hexaoxohexatungstate(2–)

Multiplicative prefixes may be used to provide alternative shorter names, *e.g.* with equivalent NbO or WO groups.

3. $[Nb_6O_{19}]^{8-}$
   μ_6-oxo-dodeca-μ-oxo-hexakis(oxoniobate)(8–)

A homopolyanion (isopolyanion) can be reduced without electron localisation. Such compounds are characterized by intervalence charge transfer absorptions in their electronic spectra and are commonly termed 'mixed valence compounds'. The above names are used with the resulting charge expressed by the charge number. For example, $[Mo_6O_{19}]^{2-}$ can be reduced to $[Mo_6O_{19}]^{3-}$ which may be named as follows:

Example:
4. $[Mo_6O_{19}]^{3-}$
   μ_6-oxo-dodeca-μ-oxo-hexakis(oxomolybdate)(3–)

However, when it is necessary to express electron localization in the reduced species, the polyanion is named in the same manner as a heterocentre polyanion. In this case, the atom with lower oxidation state comes first in the formula and precedes those with the higher valencies in the name. When numbering central atoms, the lowest possible number is assigned to the reduced atom.

II-1.3.2 **Heterocentre polyanions**

II-1.3.2.1 *Mono- or polysubstitution*

In names, central atoms are cited in alphabetical order independently of the numbering scheme. The list of central atom element names is enclosed in parentheses with the ending -ate after the parenthesis.

Examples:
1. $[NbW_5O_{19}]^{3-}$
   μ_6-oxo-dodeca-μ-oxo-hexaoxo(niobiumpentatungsten)ate(3–)

2. $[Nb_5WO_{19}]^{7-}$
   μ_6-oxo-dodeca-μ-oxo-hexaoxo(pentaniobium-1-tungsten)ate(7–), or
   μ_6-oxo-dodeca-μ-oxo-hexaoxo[pentaniobium(v)-1-tungsten(vi)]ate
   Tungsten is assigned number 1 because of its position in Table II.1.

3. $[Nb_4W_2O_{19}]^{6-}$
   μ_6-oxo-dodeca-μ-oxo-hexaoxo(tetraniobium-1,2-ditungsten)ate(6–), or
   μ_6-oxo-dodeca-μ-oxo-hexaoxo(tetraniobium-1,6-ditungsten)ate(6–)
   In this example, two isomers occur; they are commonly designated *cis* and *trans*. The numbering system outlined above coupled with coordination nomenclature provides a unique name for each isomer.

4. $[Nb_4W_2O_{19}]^{6-}$
   μ_6-oxo-dodeca-μ-oxo-hexaoxo(tetraniobiumditungsten)ate(6–)
   When the substituted positions are not known, locant designators are not given.

5. $[V_2W_4O_{19}]^{4-}$
   μ_6-oxo-dodeca-μ-oxo-hexaoxo(tetratungsten-5,6-divanadium)ate(4–), or
   μ_6-oxo-dodeca-μ-oxo-hexaoxo(tetratungsten-3,5-divanadium)ate(4–)
   Vanadium locants are related to the vanadium position in Table II.1

6. $[NbVW_4O_{19}]^{4-}$
   μ_6-oxo-dodeca-μ-oxo-hexaoxo(5-niobiumtetratungsten-3-vanadium)ate(4–)

II-1.3.2.2 *Reduced heterocentre polyanions*

Generally, the most easily reducible central atom is known.

Examples:
1. $[Nb_5WO_{19}]^{8-}$
   μ_6-oxo-dodeca-μ-oxo-hexaoxo[pentaniobium(v)tungsten(v)]ate
   In this example, tungsten is reduced. In more complicated cases, the charge number may be used.

2. $[Nb_4W_2O_{19}]^{7-}$
   μ_6-oxo-dodeca-μ-oxo-hexaoxo(tetraniobium-1,6-ditungsten)ate(7–)
   In this example, the compound has been reduced by one electron; it is a mixed valence ion. If the electron is localized (hypothetical compound), it is possible to indicate the reduced atom by using oxidation numbers,
   i.e. μ_6-oxo-dodeca-μ-oxo-hexaoxo[tetraniobium(v)-1-tungsten(v)-6-tungsten(vi)]ate.

II-1.3.3 **Heteroligand polyanions**

In the preceding compounds, metal atoms are surrounded by oxygen ligands. These can be replaced by sulfur ligands, hydroxide ligands, *etc.*

II-1.3.3.1 *Single substitution*

When one substitution takes place, abbreviated names without locants may be used.

Example:
1. $[W_6O_{18}S]^{2-}$
   μ_6-oxo-dodeca-μ-oxo-pentaoxothiohexatungstate(2–), or
   μ_6-oxo-undeca-μ-oxo-hexaoxo-μ-thio-hexatungstate(2–), or
   μ_6-oxo-dodeca-μ-oxo-pentaoxo-1a-thiohexatungstate(2–), or
   μ_6-oxo-undeca-μ-oxo-hexaoxo-1b-μ-thio-hexatungstate(2–)

II-1.3.3.2 *Several substitutions*

When two oxygen atoms of a homopolyanion (isopolyanion) are replaced, then numbering is necessary. The numbering order of central atoms may depend on the place in Table II.1 of the ligating atom of the replacing group.

Examples:

1. $[Mo_6O_{17}S_2]^{2-}$
   μ_6-oxo-dodeca-μ-oxo-tetraoxo-5d.6f-dithiohexamolybdate(2–)
   Two terminal oxygen atoms are replaced.

2. $[W_6O_{17}(OH)S]^-$
   5f-μ-hydroxo-μ_6-oxo-undeca-μ-oxo-pentaoxo-6f-thiohexatungstate(1–)
   One terminal oxygen atom is replaced by a sulfur atom, and one bridging oxygen is replaced by a hydroxo group.

3. $[NbW_5O_{18}S]^{3-}$
   μ_6-oxo-dodeca-μ-oxo-pentaoxo-1a-thio(6-niobiumpentatungsten)ate(3–)
   Sulfur is terminal and bound to tungsten.

4. $[Mo_6O_{17}(OH)_2]^{2-}$
   1e.2e-di-μ-hydroxo-μ_6-oxo-deca-μ-oxo-hexaoxo[1,2-dimolybdenum(v)-tetramolybdenum(vi)]ate
   This compound has a trivial name: molybdenum blue. It is a two-electron-reduced compound which also has fixed hydrons.

II-1.3.4 Names of more complicated species

Examples:

1. $[CeW_{10}O_{36}]^{8-}$ (Figure II-1.2)
   bis(μ_5-oxo-octa-μ-oxo-nonaoxopentatungstato)cerate(8–)
   This compound is derived from the Lindqvist structure by the loss of a tungsten atom; then two such groups are coordinated to one cerium atom.

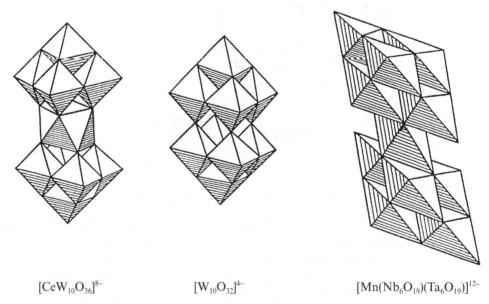

$[CeW_{10}O_{36}]^{8-}$ $[W_{10}O_{32}]^{4-}$ $[Mn(Nb_6O_{19})(Ta_6O_{19})]^{12-}$

Figure II-1.2. Examples of polyanions derived from the Lindqvist structure. $[CeW_{10}O_{36}]^{8-}$ is made of two W_5O_{18} subunits with Ce joining them, the oxygen atoms forming an antiprismatic environment; $[W_{10}O_{32}]^{4-}$ is made of two W_5O_{18} subunits directly linked; $[Mn(Nb_6O_{19})(Ta_6O_{19})]^{12-}$ is made of two Lindqvist units linked by a Mn atom, the oxygen atoms forming an octahedral environment.

2. $[W_{10}O_{32}]^{4-}$ (Figure II-1.2)

   tetra-μ-oxo-bis[μ_5-oxo-octa-μ-oxo-pentakis(oxotungstate)](4–)

   This compound has a trivial name: tungstate Y; it is made of two identical units sharing four vertices; each unit is derived from the Lindqvist structure by the loss of one tungsten atom.

3. $Mn(Nb_6O_{19})(Ta_6O_{19})]^{12-}$ (Figure II-1.2)

   [μ_6-oxo-dodeca-μ-oxo-hexakis(oxoniobato)(8–)-O^{1b}, O^{1c}, O^{2e}][μ_6-oxo-dodeca-μ-oxo--hexakis(oxotantalato)(8–)-O^{1b}, O^{1c}, O^{2e}]manganate(12–) (Note1b)

   This compound contains two Lindqvist units; each is linked by three oxygens ligated to manganese which is surrounded by an octahedron of oxygen ligands.

II-1.4 POLYANIONS WITH THE ANDERSON STRUCTURE

II-1.4.1 Polyanions with seven central atoms

These compounds are derived from $[NH_4]_6[Mo_6TeO_{24}]$, the structure of which was proposed by Anderson and determined by Evans. The seven atoms of tellurium and molybdenum are all octahedrally surrounded and in the same plane, or nearly so (Figure II-1.3). These polyanions have D_{3d} symmetry. Examples are $[Mo_6TeO_{24}]^{6-}$, $[IMo_6O_{24}]^{5-}$, $[H_6CrMo_6O_{24}]^{3-}$.

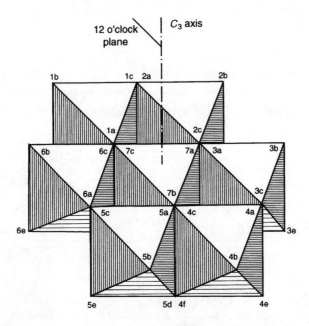

Figure II-1.3. Octahedron assembly and locant designators in the Anderson structure.

The central atoms, all octahedrally surrounded, are considered equivalent and their names and symbols are cited in alphabetical order in names and formulae. The numbering of these structures is achieved by starting with the peripheral octahedra, termed the crown, wherever the central atom of the crown occurs in Table II-1. Thus, the structural analogy between all of these compounds and those where the central atom located at the centre of the crown can be removed and replaced by two tetrahedra, one on each side of the crown, is clearly indicated (see II-1.4.2).

Note 1b. Where a name is divided over two or more lines of text, a hyphen which is an integral part of that name and which occurs at the line break is repeated at the beginning of the succeeding line.

Examples:

1. $[Mo_6TeO_{24}]^{6-}$
   hexa-μ_3-oxo-hexa-μ-oxo-dodecaoxo(hexamolybdenum-7-tellurium)ate(6–)

2. $[H_6CrMo_6O_{24}]^{3-}$
   hexa-μ_3-hydroxo-hexa-μ-oxo-dodecaoxo[7-chromium(III)hexamolybdenum]ate(3–)
   In this example hydrogen positions are not known, but it is assumed they are part of the hydroxo groups on the six μ_3 bridges. If the hydrogen positions are not to be shown, the name should be hexahydrogenhexa-μ_3-oxo-hexa-μ-oxo-dodecaoxo[7--chromium(III)hexamolybdenum]ate(3–).
   An isomer of this polyanion, $[H_6CrMo_6O_{24}]^{3-}$, has been postulated with the chromium atom in the peripheral crown.

3. $[H_6CrMo_6O_{24}]^{3-}$
   1b.1e-diaqua-1a.1f-di-μ_3-hydroxo-1c.1d-di-μ-hydroxo-2c.3f,4c.5f-tetra-μ_3-oxo-2f.3c,4f.5c--tetra-μ-oxo-decaoxo[1-chromium(III)hexamolybdenum]ate

II-1.4.2 Names of more complicated species

There are several examples of polyanions derived from the Anderson-type polyanion. Such compounds usually have D_{3d} symmetry. The structure comprises six octahedra fused in a crown plus two tetrahedra attached by their base on each side of the crown. These tetrahedra share three oxygen atoms with the six octahedra.

Examples:

1. $[As_2Mo_6O_{26}]^{6-}$
   hexa-μ-oxo-bis-μ_6-(tetraoxoarsenato-O,O',O'')-hexakis(dioxomolybdate)(6–)

2. $[Mo_8O_{26}]^{4-}$
   hexa-μ-oxo-bis-μ_6-(tetraoxomolybdato-O,O',O'')-hexakis(dioxomolybdate)(4–)

3. $[Mo_6O_{24}(AsC_6H_5)_2]^{4-}$
   hexa-μ-oxo-bis-μ_6-(phenylarsonato-O,O',O'')-hexakis(dioxomolybdate)(4–)

Since there are two different environments — one tetrahedral and one octahedral — and since the tetrahedral group can be replaced, it is useful to distinguish the different coordination geometries. This is accomplished by treating the tetrahedron as a ligand. Another advantage of this particular treatment of tetrahedra will be seen later in the description of Keggin-type polyanions. Thus, for Anderson structures, the numbering of the central atoms in the crown is in no case altered by changing the tetrahedral bridging groups.

II-1.5 POLYANIONS WITH TWELVE CENTRAL ATOMS

The structure of the anion $[PW_{12}O_{40}]^{3-}$ is known as the Keggin structure. Many compounds have this or a closely related structure. In such formulae the heteroatom is cited first in order to conform with common usage in heteropolyanion chemistry. If there are several different metal atoms, they are given in alphabetical order. Finally, ligand symbols are cited as in coordination compound formulae.

II-1.5.1 Compounds with the Keggin structure and isomers

II-1.5.1.1 Compounds containing only one kind of transition metal

This family is characterized by the general formula $[XM_{12}O_{40}]^{n-}$, for example $[SiW_{12}O_{40}]^{4-}$ or $[PMo_{12}O_{40}]^{3-}$. M can be either tungsten or molybdenum; X can be silicon, germanium, phosphorus, arsenic, or boron.

The polyanion is made of four M_3O_{13} groups which share vertices. Such a group has a trigonal axis of symmetry and contains three octahedra sharing edges. The three octahedra have a common vertex which is also a vertex of the central XO_4 tetrahedron. The central XO_4 group is treated as a bridging ligand. The structure has T_d symmetry (Figures II-1.4 and II-1.5). In some cases, X may be absent.

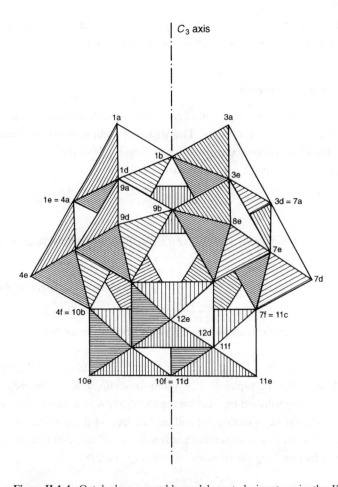

Figure II-1.4. Octahedron assembly and locant designators in the Keggin structure (isomer 1). This isomer, sometimes trivially called α, has T_d symmetry.

Example:
1. $[SiW_{12}O_{40}]^{4-}$
   1c.2b,1b.3c,1e.4a,1d.9a,2c.3b,2d.5a,2e.6a,3d.7a,3e.8a,4c.5b,4d.9e,4f.10b,5e.6d,5f.10c, 6c.7b,6f.11b,7e.8d,7f.11c,8c.9b,8f.12b,9f.12c,10f.11d,10d.12f,11f.12d-tetracosa-μ-oxo--μ_{12}-(tetraoxosilicato-$O^{1.4.9},O^{2.5.6},O^{3.7.8},O^{10.11.12}$)-dodecakis(oxotungstate)(4−)

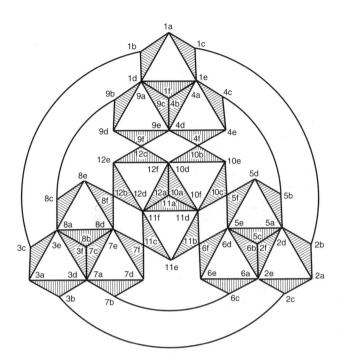

Figure II-1.5. Opened two-dimensional Keggin structure; the curved lines join octahedron vertices which are shared in the actual three-dimensional structure (isomer 1).

Five isomers are likely to occur; in the above structure, one or several M_3O_{13} groups can be rotated by 60° around their threefold axes. This brings additional difficulty to the nomenclature of these polyanions. Two structures are presently known as α and β. The former (α) has the Keggin structure, *i.e.* T_d symmetry. The latter (β) is derived by rotating one M_3O_{13} group (Figures II-1.6 and II-1.7). It has C_{3v} symmetry. The third known isomer is illustrated by $[Al_{13}O_4(OH)_{24}(H_2O)_{12}]^{7+}$ wherein the central group is AlO_4. In this compound all four groups are rotated as compared to the α isomer; the symmetry is T_d. The numbering of octahedral central atoms gives an unambiguous name.

Example:
2. $[SiW_{12}O_{40}]^{4-}$; trivially called β (Figure II-1.6)
   1c.2b,1b.3c,1e.4a,1d.9a,2c.3b,2d.5a,2e.6a,3d.7a,3e.8a,4c.5b,4d.9e,4f.10c,5e.6d,5f.11b, 6c.7b,6f.11c,7e.8d,7f.12b,8c.9b,8f.12c,9f.10b,10f.11d,10d.12f,11f.12d-tetracosa-μ-oxo-μ_{12}- -(tetraoxosilicato-$O^{1,4,9}, O^{2,5,6}, O^{3,7,8}, O^{10,11,12}$)-dodecakis(oxotungstate)(4–)

In order to keep names as simple as possible, only oxygen atoms bridging skeletal planes need to be designated. These oxygen atoms are given the set of two numbers which refer to the central atoms joined by the considered oxygen atoms. This results in the following sequence of numbers for the bridging oxygens which quite clearly differentiate the five isomers (Figure II-1.7):

- the basic Keggin structure: α isomer 1
  1.4,1.9,2.5,2.6,3.7,3.8,4.10,5.10,6.11,7.11,8.12,9.12.
- one M_3O_{13} group is rotated by 60°: β isomer 2
  1.4,1.9,2.5,2.6,3.7,3.8,4.10,5.11,6.11,7.12,8.12,9.10.

POLYANIONS

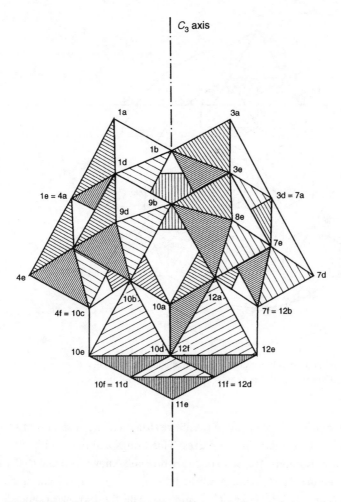

Figure II-1.6. Octahedron assembly and locant designators in isomer 2, which has C_{3v} symmetry, derived from isomer 1 (Keggin structure, Fig. II-1.7) by rotating one W_3O_{13} group by 60°. Trivially this isomer is called β.

- two M_3O_{13} groups are rotated by 60°: γ isomer 3
  1.3,1.6,2.4,2.5,3.7,3.9,4.7,4.10,5.8,5.11,6.8,6.12,7.9,7.10,8.11,8.12.
- three M_3O_{13} groups are rotated by 60°: δ isomer 4
  1.4,1.6,2.4,2.5,3.5,3.6,4.7,4.8,5.9,5.10,6.11,6.12.
- four M_3O_{13} groups are rotated by 60°: $[Al_{13}H_{48}O_{40}]^{7+}$, ε isomer 5
  1.4,1.4,2.5,2.5,3.6,3.6,4.7,4.12,5.8,5.9,6.10,6.11.

Examples:

3. $[SiW_{12}O_{40}]^{4-}$, Keggin structure
   1.4,1.9,2.5,2.6,3.7,3.8,4.10,5.10,6.11,7.11,8.12,9.12-dodeca-μ-oxo-μ_{12}-(tetraoxosilicato-$O^{1.4.9}, O^{2.5.6}, O^{3.7.8}, O^{10.11.12}$)-tetrakis[tri-$\mu$-oxo-tris(oxotungstate)](4−)

4. $[SiW_{12}O_{40}]^{4-}$, one M_3O_3 group rotated
   1.4,1.9,2.5,2.6,3.7,3.8,4.10,5.11,6.11,7.12,8.12,9.10-dodeca-μ-oxo-μ_{12}-(tetraoxosilicato-$O^{1.4.9}, O^{2.5.6}, O^{3.7.8}, O^{10.11.12}$)-tetrakis[tri-$\mu$-oxo-tris(oxotungstate)](4−)

5. $[Al_{13}H_{48}O_{40}]^{7+}$ (Figure II-1.7)
   1.4,1.4,2.5,2.5,3.6,3.6,4.7,4.12,5.8,5.9,6.10,6.11-dodeca-μ-hydroxo-μ_{12}-
   -(tetraoxoaluminato-$O^{1.2.3}, O^{4.7.12}, O^{5.8.9}, O^{6.10.11}$)-tetrakis[tri-$\mu$-hydroxo-
   -tris(aquaaluminium)](7+)

Trivial nomenclature has developed in an anarchic manner for polytungstates. For example, metatungstate is a dodecatungstate of the same structure as α-$[SiW_{12}O_{40}]^{4-}$ in which two hydrons are trapped in the central cavity in place of the silicon atom. Polytungstate X is the β isomer of the preceding metatungstate while tungstate Y is a decatungstate derived from the Lindqvist structure (see Section II-1.3.4).

Examples:

6. $[H_2W_{12}O_{40}]^{6-}$, metatungstate
   di-μ_3-hydroxo-1.4,1.9,2.5,2.6,3.7,3.8,4.10,5.10,6.11,7.11,8.12,9.12-di-μ_3-oxo-dodeca-μ-
   -oxo-tetrakis[tri-μ-oxo-tris(oxotungstate)](6−)

7. $[H_2W_{12}O_{40}]^{6-}$, polytungstate X
   di-μ_3-hydroxo-1.4,1.9,2.5,2.6,3.7,3.8,4.10,5.11,6.11,7.12,8.12,9.10-di-μ_3-oxo-dodeca-μ-
   -oxo-tetrakis[tri-μ-oxo-tris(oxotungstate)](6−)

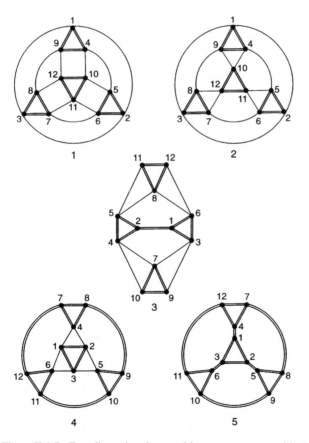

Figure II-1.7. Two-dimensional opened isomer structures with the formula $XW_{12}O_{40}$. Black points are central atoms. A single line between two central atoms means a common octahedral vertex. A double line between two central atoms means a shared octahedral edge. Numbers under the structures refer to the isomer number in the text.

II-1.5.1.2   *Compounds with several transition metals, i.e. substituted compounds*

This class contains many more compounds than the preceding class. Indeed, the tetrahedrally-surrounded atom can be substituted as well as one or several tungsten atoms. For instance, the following species (ignoring the charge) are known with the following substitutions:

$SiW_{11}ZO_{40}$ 	$Z = Cu^{II}, Zn^{II}, Mn^{II}, Co^{II}, Ni^{II}, Fe^{II}, Mn^{III}, Co^{III}, Fe^{III}, Cr^{III}, Al^{III}$

$SiW_{10}Z_2O_{40}$ or $SiW_{10}ZZ'O_{40}$ 	$Z = Mo^{VI}, V^V, V^{IV}$; $Z' = Mo^{VI}, V^V, V^{IV}$

$SiW_9Z_3O_{40}$, $SiW_9Z_2Z'O_{40}$ or $SiW_9ZZ'Z''O_{40}$ 	$Z = Z' = Z'' = Mo^{VI}, V^V, V^{IV}, Co^{II}, Mn^{III}, Fe^{III}$

(a) *Monosubstituted Compounds.* The choice of the reference axis and the reference plane is governed by the recommendations given earlier; the hierarchy is: (i) the symmetry of the idealised polyanion framework; (ii) the position of the substituting atom in Table II-1 with respect to other central atoms.

Examples:
1. $[SiMoW_{11}O_{40}]^{4-}$, trivially referred to as α
   1.4,1.9,2.5,2.6,3.7,3.8,4.10,5.10,6.11,7.11,8.12,9.12-tetracosa-μ-oxo-dodecaoxo-μ_{12}-
   -(tetraoxosilicato-$O^{1.4.9}, O^{2.5.6}, O^{3.7.8}, O^{10.11.12}$)-(1-molybdenumundecatungsten)ate(4–)

2. $[SiCoW_{11}O_{39}(H_2O)]^{6-}$, trivially referred to as β
   1a-aqua-1.4,1.9,2.5,2.6,3.7,3.8,4.10,5.11,6.11,7.12,8.12,9.10-tetracosa-μ-oxo-undecaoxo-
   -μ_{12}-(tetraoxosilicato-$O^{1.4.9}, O^{2.5.6}, O^{3.7.8}, O^{10.11.12}$)-(1-cobaltundecatungsten)ate(6–)

3. $[SiNbW_{11}O_{40}]^{5-}$, trivially referred to as α
   1.4,1.9,2.5,2.6,3.7,3.8,4.10,5.10,6.11,7.11,8.12,9.12-tetracosa-μ-oxo-dodecaoxo-μ_{12}-
   -(tetraoxosilicato-$O^{1.4.9}, O^{2.5.6}, O^{3.7.8}, O^{10.11.12}$)-(12-niobiumundecatungsten)ate(5–)

For the β structure, three isomers are likely to occur. They are trivially designated as β_1, β_2, and β_3 depending on the position of the substituted atom, *i.e.* whether it is located on the farthest position to the rotated group, on an adjacent position to the rotated group, or on an atom of the rotated group, respectively.

Examples:
4. $[SiVW_{11}O_{40}]^{5-}$, trivially referred to as β_1
   1.4,1.9,2.5,2.6,3.7,3.8,4.10,5.11,6.11,7.12,8.12,9.10-tetracosa-μ-oxo-dodecaoxo-μ_{12}-
   -(tetraoxosilicato-$O^{1.4.9}, O^{2.5.6}, O^{3.7.8}, O^{10.11.12}$)-(undecatungsten-3-vanadium)ate(5–)
   In this example, the reference axis is, of course, the C_3 axis which is the axis around which the M_3O_{13} group has turned. The preferred terminal skeletal plane is the less condensed one, even though it contains the vanadium atom which comes after tungsten in Table II-1 [see Section II-1.2.4(b)]. The reference symmetry plane is chosen to give the lowest locants to tungsten, *i.e.* so that the number 3 is assigned to the vanadium.

5. $[SiVW_{11}O_{40}]^{5-}$, trivially referred to as β_3
   1.4,1.9,2.5,2.6,3.7,3.8,4.10,5.11,6.11,7.12,8.12,9.10-tetracosa-μ-oxo-dodecaoxo-μ_{12}-
   -(tetraoxosilicato-$O^{1.4.9}, O^{2.5.6}, O^{3.7.8}, O^{10.11.12}$)-(undecatungsten-12-vanadium)ate(5–)

6. $[SiVW_{11}O_{40}]^{5-}$, trivially referred to as β_2
   1.4,1.9,2.5,2.6,3.7,3.8,4.10,5.11,6.11,7.12,8.12,9.10-tetracosa-μ-oxo-dodecaoxo-μ_{12}-
   -(tetraoxosilicato-$O^{1.4.9}, O^{2.5.6}, O^{3.7.8}, O^{10.11.12}$)-(undecatungsten-9-vanadium)ate(5–)
   Enantiomeric structures occur for this compound.

Recommendations can be applied to larger compounds. For instance $[SiFeW_{11}O_{39}(OH)]^{6-}$ can dimerize yielding $[SiW_{11}O_{39}Fe-O-FeSiW_{11}O_{39}]^{12-}$, each FeW_{11} unit of which has T_d symmetry.

Example:

7. $[SiW_{11}O_{39}FeOFeSiW_{11}O_{39}]^{12-}$
   1a-μ-oxo-bis[1.4,1.9,2.5,2.6,3.7,3.8,4.10,5.10,6.11,7.11,8.12,9.12-tetracosa-μ-oxo-
   -undecaoxo-μ_{12}-(tetraoxosilicato-$O^{1.4.9}, O^{2.5.6}, O^{3.7.8}, O^{10.11.12}$)-(1-ironundecatungsten)]-
   -ate(12–)

(b) *Disubstituted and polysubstituted compounds.* The numbering system provides a solution to the problem of distinguishing isomers of di- and poly-substituted compounds. For example $[PMo_{10}V_2O_{40}]^{5-}$ has five isomers. Because of its position in Table II-1, one vanadium must be given the locant 12 in the metal framework with T_d symmetry. The isomers are now distinguished by the locant numbers of vanadium:

12 and 11   12 and 8   12 and 7   12 and 3   12 and 2.

A trisubstituted compound containing two molybdenum atoms and one vanadium atom has been prepared. The locant 1 is given to the atom coming first in Table II-1, *i.e.* molybdenum, and a locant number as high as possible is then given to the atom appearing last, *i.e.* vanadium.

Example:

8. $[SiMo_2VW_9O_{40}]^{5-}$
   1.4,1.9,2.5,2.6,3.7,3.8,4.10,5.10,6.11,7.11,8.12,9.12-tetracosa-μ-oxo-dodecaoxo-μ_{12}-
   -(tetraoxosilicato-$O^{1.4.9}, O^{2.5.6}, O^{3.7.8}, O^{10.11.12}$)-[1,2-dimolybdenum(VI)nonatungsten(VI)-3-
   -vanadium(V)]ate(5–)

II-1.5.1.3 Ligand substitution

When a ligand is substituted, the same procedure can be applied using locant designators. The anions $[HW_{12}F_2O_{38}]^{5-}$, $[H_2W_{12}F_2O_{38}]^{4-}$, and $[HW_{12}F_3O_{37}]^{4-}$ are examples where, in every case, the hydrons are trapped in the central cavity, and where fluorine atoms always bridge three tungsten atoms.

Examples:

1. $[HW_{12}F_2O_{38}]^{5-}$
   μ_3-fluoro-μ_3-(hydrogenfluoride)-1.4,1.9,2.5,2.6,3.7,3.8,4.10,5.10,6.11,-7.11,8.12,9.12-di-
   -μ_3-oxo-dodeca-μ-oxo-tetrakis[tri-μ-oxo-tris(oxotungstate)](5–)

II-1.5.1.4 Reduced compounds

A reduced compound is treated in the same manner as a substituted compound if the added electrons are localized. However, in many cases, the hopping process makes localization impossible. Then only the overall charge is changed.

Examples:

1. $[SiMo_{12}O_{40}]^{4-}$ can be reduced by four electrons to give $[SiMo_{12}O_{40}]^{8-}$
   1.4,1.9,2.5,2.6,3.7,3.8,4.10,5.10,6.11,7.11,8.12,9.12-dodeca-μ-oxo-μ_{12}-(tetraoxosilicato-$O^{1.4.9},O^{2.5.6},O^{3.7.8},O^{10.11.12}$)-tetrakis(tri-$\mu$-oxo-tris(oxomolybdate)](8–)

2. When electrons are added to $[SiMo_{12}O_{40}]^{4-}$, to produce the doubly reduced $[SiMo_{12}O_{40}]^{6-}$, they are assumed to be localized on two molybdenum atoms.
   This is also assumed in $[SiMo_2W_{10}O_{40}]^{6-}$:
   1.4,1.9,2.5,2.6,3.7,3.8,4.10,5.10,6.11,7.11,8.12,9.12-tetracosa-μ-oxo-dodecaoxo-μ_{12}-(tetraoxosilicato-$O^{1.4.9},O^{2.5.6},O^{3.7.8},O^{10.11.12}$)-[1,2-dimolybdenum(v)decatungsten(vi)]ate(6–)

II-1.5.2 Compounds in which central atoms are missing (defect structures)

Several polyanions have been prepared which can be described from the Keggin structure by removing tungsten atoms and their associated non-bridging oxygen atoms.

II-1.5.2.1 Compounds with one vacancy

For numbering central atoms, the recommendations given earlier are used: the upper skeletal plane is the plane with the lowest number of central atoms, *i.e.* the one with the vacant position, and with the less condensed octahedra.

Herein several possibilities can occur because there are three ways to locate the vacant position with respect to the 12 o'clock position. With the upper skeletal plane consisting of only tungsten atoms, the three possibilities are represented in the following diagram which shows the uppermost skeletal plane.

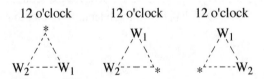

The asterisk represents the vacant site. The rule to locate the vacancy is: 'the vacant site is considered as an atom position and then numbered; it is given the lowest possible number'. Then the choice is:

Examples:

1. $[SiW_{11}O_{39}]^{8-}$ (derived from the Keggin structure)
   1c.2b,1d.4a,1e.5a,2d.6a,2e.7a,3c.4b,3d.8e,3f.9b,4e.5d,4f.9c,5c.6b,5f.10b,6e.7d,6f.10c,
   7c.8b,7f.11b,8f.11c,9f.10d,9d.11f,10f.11d-icosa-μ-oxo-pentadecaoxo-μ_{11}-
   -(tetraoxosilicato-$O^{1.4.5},O^{2.6.7},O^{3.8},O^{9.10.11}$)-undecatungstate(8–)

   From the C_{3v} isomer derived from the Keggin structure, three isomers can be obtained by removing one central atom. The choice of the reference axis takes precedence over any other choice, *i.e.* wherever the vacancy is.

In the following example, the central atom is removed from the rotated M_3O_{13} group.

2. $[SiW_{11}O_{39}]^{8-}$
   1c.2a,1e.4a,1f.5a,2e.6a,2f.7a,3e.4d,3b.8c,3f.9c,4c.5b,4f.10b,5e.6d,5f.10b,6c.7b,6f.11b,
   7e.8d,7f.11c,8f.9b,9e.10d,9d.11e,10e.11d-icosa-μ-oxo-pentadecaoxo-μ_{11}-(tetraoxosilicato-
   -$O^{1.2},O^{3.8.9},O^{4.5.10},O^{6.7.11}$)-undecatungstate(8–)

In the following example, the central atom is removed from the plane containing six central atoms. From the rule above, the 12 o'clock position is defined with the vacant position. There are two optical isomers of this structure. As in the general practice, they have the same name; the numbering indicates chirality clockwise or anticlockwise depending on the isomer.

3. $[SiW_{11}O_{39}]^{8-}$
   1b.3c,1c.2b,1d.8a,2c.3b,2d.4a,2e.5a,3d.6a,3e.7a,4e.5d,4f.10b,5c.6b,5f.10c,6e.7d,6f.11b,
   7c.8b,7f.11c,8f.9b,9d.11f,9f.10d,10f.11d-icosa-μ-oxo-pentadecaoxo-μ_{11}-(tetraoxosilicato-
   -$O^{1.8},O^{2.4.5},O^{3.6.7},O^{9.10.11}$)-undecatungstate(8–)

   In this example, the central atom is removed from the plane opposite the plane of the rotated group.

4. $[SiW_{11}O_{39}]^{8-}$
   1c.2b,1d.4a,1e.5a,2d.6a,2e.7a,3c.4b,3d.8e,3f.9c,4e.5d,4f.10b,5c.6b,5f.10c,6e.7d,6f.11b,
   7c.8b,7f.11c,8f.9b,9f.10d,10f.11d,9d.11f-isoca-μ-oxo-pentadecaoxo-μ_{11}-(tetraoxosilicato-
   -$O^{1.4.5},O^{2.6.7},O^{3.8},-O^{9.10.11}$)-undecatungstate(8–)

II-1.5.2.2   *Compounds with three vacancies*

Two types are known: in the first, a complete W_3O_{13} group has been removed (trivially named B type) while in the second, three adjacent metal atoms belonging to three different W_3O_{13} groups have been removed (trivially named A type). In each case one remaining M_3O_{13} group can be rotated by 60°.

Examples:

1. $[PMo_9O_{28}(OH)_6]^{3-}$, type A derived from T_d Keggin structure (Figure II-1.8,I)
   4f,5f,6f,7f,8f,9f-hexahydroxo-1c.2a,1a.3c,1f.4a,1e.9a,2c.3a,2e.5a,2f.6a,3e.7a,3f.8a,4c.5b,
   4d.9e,5e.6d,6b.7b,7e.8d,8c.9b-pentadeca-μ-oxo-μ_9-[tetraoxophosphato(v)-$O^{1.2.3},O^{4.5}$,
   -$O^{6.7}O^{8.9}$]-nonakis(oxomolybdate)(3–)

2. $[As^{III}Mo_9O_{33}]^{9-}$, type B derived from C_{3v} Keggin structure (Figure II-1.8,II)

1c.2a,1d.3a,2d.3c,2b.4b,2f.7b,3e.5a,3f.8a,4e.6c,4f.7c,5e.8c,5f.9c,6f.7e,6d.9d,7d.8d,8f.9e--pentadeca-μ-oxo-pentadecaoxo-μ_9-[trioxoarsenato(III)-$O^{1.2.3},O^{4.6.7},O^{5.8.9}$]--nonamolybdate(9–)

This isomer is, as yet, unknown.

3. [As^{III}W$_9$O$_{33}$]$^{9-}$, type B derived from T_d Keggin structure (Figure II-1.8,III)
1c.2b,1b.3c,1e.4a,1d.9a,2c.3b,2d.5a,2e.6a,3d.7a,3e.8a,4c.5d,4d.9e,5e.6d,6c.7b,7e.8d, 8c.9b-pentadeca-μ-oxo-pentadecaoxo-μ_9-[trioxoarsenato(III)-$O^{1.4.9},O^{2.5.6},O^{3.7.8}$]--nonatungstate(9–)

4. [SiW$_9$O$_{34}$]$^{10-}$, type A derived from C_{3v} Keggin structure (Figure II-1.8,IV)
1c.2a,1a.3c,1f.4a,1e.9a,2c.3a,2e.5a,2f.6a,3e.7a,3f.8a,4e.5d,4b.9c,5c.6b,6e.7d,7c.8b,8e.9d--pentadeca-μ-oxo-pentadecaoxo-μ_9-(tetraoxosilicato-$O^{1.2.3},O^{4.9},O^{5.6},O^{7.8}$)--nonatungstate(10–)

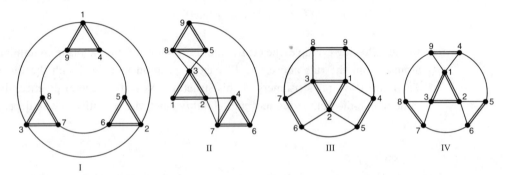

Figure II-1.8. Two-dimensional opened structures with formula XM$_9$O$_n$. Black points are central atoms. A single line refers to a shared octahedral vertex. A double line refers to a shared octahedral edge.

POLYANIONS WITH EIGHTEEN CENTRAL ATOMS

Another class of polyanions contains eighteen central atoms. The basic structure is known as the Dawson structure. It has D_{3h} symmetry (Figure II-1.9). Examples are [P$_2$W$_{18}$O$_{62}$]$^{6-}$, [P$_2$Mo$_{18}$O$_{62}$]$^{6-}$, [As$_2$W$_{18}$O$_{62}$]$^{6-}$, and [As$_2$Mo$_{18}$O$_{62}$]$^{6-}$.

These compounds are dimers of XM$_9$ which are derived from the Keggin structure XM$_{12}$. Three tungsten atoms are removed, each from a different M$_3$O$_{13}$ group. The XM$_9$ units are joined together by sharing octahedral vertices. The internal XO$_4$ tetrahedron has a basal plane towards the open side of the XW$_9$ moiety.

The X$_2$M$_{18}$ compound can be named as a complete unit using recommendations given earlier. The upper skeletal plane is an M$_3$O$_{13}$ group. Since there are two identical terminal planes, the upper one and/or the lower one, and since such a group may be rotated by 60°, three isomers are likely to occur; among them, two are known:

Isomer 1	Isomer 2	Isomer 3
uppermost skeletal plane not rotated; lowest skeletal plane not rotated.	uppermost skeletal plane not rotated; lowest skeletal plane rotated	uppermost skeletal plane rotated; lowest skeletal plane rotated

POLYANIONS

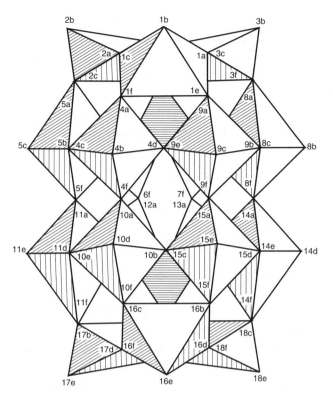

Figure II-1.9. Assembly of the octahedra and locant designators for the ion $X_2M_{18}O_{62}$ comprised of two XM_9 subunits.

For these three cases, the locant sequences of the μ-oxo bridges are as follows:

Isomer 1: 1c.2a,1a.3c,1f.4a,1e.9a,2c.3a,2e.5a,2f.6a,3e.7a,3f.8a,**4c.5b**,4d.9e,4f.10a,5e.6d, 5f.11a,6c.7b,6f.12a,7e.8d,7f.13a,8c.9b,8f.14a,9f.15a,10e.11d,10b.15c,<u>10f.16c</u>,11c.12b,11f.17b, 12e.13d,12f.17c,13c.14b,13f.18b,14e.15d,16f.18c,15f.16b,16f.17d,16d.18f,17f.18d.

Isomer 2: 1c.2a,1a.3c,1f.4a,1e.9a,2c.3a,2e.5a,2f.6a,3e.7a,3f.8a,**4e.5d**,4b.9c,4f.10a,5c.6b, 5f.11a,6e.7d,6f.12a,7c.8b,7f.13a,8e.9d,8f.14a,9f.15a,10c.11b,10d.15e,<u>10f.16b</u>,11e.12d,11f.16c, 12c.13b,12f.17b,13e.14d,13f.17c,14c.15b,14f.18b,15f.18c,16f.17d,16d.18f,17f.18d.

Isomer 3: 1c.2a,1a.3c,1f.4a,1e.9a,2c.3a,2e.5a,2f.6a,3e.7a,3f.8a,**4e.5d**,4b.9c,4f.10a,5c.6b, 5f.11a,6e.7d,6f.12a,7c.8b,7f.13a,8e.9d,8f.14a,9f.15a,10c.11b,10d.15e,<u>10f.16c</u>,11e.12d,11d.17b, 12c.13b,12f.17c,13e.14d,13f.18b,14c.15b,14f.18c,15f.16b,16f.17d,16d.18f,17f.18d.

The bold locants are the first to distinguish isomer 2 and isomer 3 from isomer 1; the underlined locants are the first to distinguish isomer 3 and isomer 1 from isomer 2.

Examples:

1. $[P_2W_{18}O_{62}]^{6-}$ (isomer 1)
   1c.2a,1a.3c,1f.4a,1e.9a,2c.3a,2e.5a,2f.6a,3e.7a,3f.8a,4c.5b,4d.9e,4f.10a,5e.6d,5f.11a,6c.7b, 6f.12a,7e.8d,7f.13a,8c.9b,8f.14a,9f.15a,10e.11d,10b.15c,10f.16c,11c.12b,11f.17b,12e.13d, 12f.17c,13c.14b,13f.18b,14e.15d,16f.18c,15f.16b,16f.17d,16d.18f,17f.18d-hexatriaconta-μ- -oxo-μ_9-(tetraoxophosphato-$O^{1,2,3}$,$O^{4,5}$,$O^{6,7}$,$O^{8,9}$)-μ_9-(tetraoxophosphato-$O^{10,11}$,$O^{12,13}$,$O^{14,15}$, $O^{16,17,18}$)-octadecakis(oxotungstate)(6–)

In order to have more compact names, multiplicative prefixes can be used when the polyanion is symmetrical. It must be pointed out that, since the polyanion has a symmetry plane perpendicular to the reference axis, the numbering is valid only for the first moiety.

2. $[P_2W_{18}O_{62}]^{6-}$ (isomer 1)
4.10,5.11,6.12,7.13,8.14,9.15-hexa-μ-oxo-bis[1c.2a,1a.3c,1f.4a,1e.9a,2c.3a,2e.5a,2f.6a, 3e.7a,3f.8a,4c.5b,4d.9e,5e.6d,6c.7b,7e.8d,8c.9b-pentadeca-μ-oxo-μ_9-(tetraoxophosphato--$O^{1,2,3}, O^{4,5}, O^{6,7}, O^{8,9}$)-nonakis(oxotungstate)](6–)

II-1.7 CONCLUSION

The names given to polyanions are complicated and carry a long sequence of numbers and letters. However, these names inevitably arise from the structural complexity of the polyanions themselves. However, their use in everyday practice may become tedious, and if a shorter name is desired, it is possible to consult Chapter I-9 on 'Oxoacids and Derived Anions' in Note 1a. The present Chapter provides a method for the naming of an oxoanion of a given formula but an unknown structure. It is commonly accepted that an abbreviated name may be used, provided that it is clearly defined at the outset, and the fully systematic name need only be given once *in extenso* at the appropriate point in the text.

The names provided in this Chapter have the merit of being systematic. The recommendations are intended to deal with all possible developments in this area.

II-2 Isotopically Modified Inorganic Compounds

CONTENTS

II-2.1 Introduction
II-2.2 Classification and symbolism
   II-2.2.1 Isotopically unmodified compounds
   II-2.2.2 Isotopically modified compounds
   II-2.2.3 Nuclide symbols
   II-2.2.4 Names for hydrogen atoms, ions or groups
   II-2.2.5 Order of nuclide symbols
II-2.3 Isotopically substituted compounds
   II-2.3.1 Definition of isotopically substituted compounds
   II-2.3.2 Formulae of isotopically substituted compounds
   II-2.3.3 Names of isotopically substituted compounds
II-2.4 Isotopically labelled compounds
   II-2.4.1 Definition of isotopically labelled compounds
   II-2.4.2 Specifically labelled compounds
     II-2.4.2.1 Definition of specifically labelled compounds
     II-2.4.2.2 Formulae of specifically labelled compounds
     II-2.4.2.3 Names of specifically labelled compounds
     II-2.4.2.4 Types of specifically labelled compounds
   II-2.4.3 Selectively labelled compounds
     II-2.4.3.1 Definition of selectively labelled compounds
     II-2.4.3.2 Formulae of selectively labelled compounds
     II-2.4.3.3 Names of selectively labelled compounds
     II-2.4.3.4 Indication of the number of labelling nuclides
     II-2.4.3.5 General labelling
     II-2.4.3.6 Uniform labelling
   II-2.4.4 Non-selectively labelled compounds
     II-2.4.4.1 Definition of non-selectively labelled compounds
     II-2.4.4.2 Isotopically labelled non-molecular materials
   II-2.4.5 Isotopically deficient compounds
II-2.5 Locants and numbering of isotopically modified compounds
   II-2.5.1 Indication of the position of isotopic modifications
   II-2.5.2 Priority
   II-2.5.3 Numbering
   II-2.5.4 Group symbols and prefixes
   II-2.5.5 Italicized nuclide symbols
II-2.6 Summary of types of isotopically modified compounds

ISOTOPICALLY MODIFIED INORGANIC COMPOUNDS

II-2.1 INTRODUCTION

These recommendations provide a general system of nomenclature for inorganic compounds whose isotopic nuclide composition (Note 2a) deviates from that which occurs in Nature (Note 2b). (In these recommendations, the term compound includes ions, radicals, and other species). They are also suitable for designating individual isotopic molecular species. An isotopically modified organic residue occurring in an organic compound, such as an organic ligand in a coordination entity, is named according to the recommendations given in the Nomenclature of Organic Chemistry, Isotopically Modified Compounds (Note 2c).

II-2.2 CLASSIFICATION AND SYMBOLISM·

II-2.2.1 Isotopically unmodified compounds

An isotopically unmodified compound has a macroscopic composition such that its constituent nuclides are present in the proportion occurring in Nature. Its formula and name are written according to IUPAC recommendations (Note 2d). The name of an isotopically unmodified compound does not require alteration unless it is desired to contrast the natural compound to the isotopically modified compound or to emphasise its natural or normal character.

Examples:
1. NH_3 — ammonia
2. Na_2CO_3 — sodium carbonate
3. HNO_3 — nitric acid
4. $K[AgF_4]$ — potassium tetrafluoroargentate
5. PH_3 — unmodified phosphane or phosphine (see Note 2d)
6. I_2O_5 — unmodified diiodine pentaoxide

II-2.2.2 Isotopically modified compounds (Note 2e)

An isotopically modified compound has a macroscopic composition such that the isotopic ratio of nuclides for at least one element deviates measurably from that which occurs in Nature.

Isotopically modified compounds may be classified into the two broad categories: isotopically substituted compounds (see Section II-2.3), and isotopically labelled compounds (see Section II-2.4).

Note 2a. *Quantities, Units and Symbols in Physical Chemistry*, 2nd Edition, Blackwell Scientific Publications, Oxford 1993.
Note 2b. For discussion of the meaning of natural composition see Atomic Weights of the Elements 1979, *Pure Appl. Chem.*, **52**, 2349 (1980).
Note 2c. *Nomenclature of Organic Chemistry*, 1979 Edition, Pergamon Press, Oxford, Section H.
Note 2d. *Nomenclature of Inorganic Chemistry, Recommendations 1990*, Blackwell Scientific Publications, Oxford, 1990.
Note 2e. A summary of the various types of isotopic modification is given in Section II-2.6.

II-2.2.3 Nuclide symbols

The symbol for denoting a nuclide in a formula or name of an isotopically modified compound consists of the atomic symbol for the element and an arabic number in the left superscript position which indicates the mass number of the nuclide (Note 2d). A metastable nuclide is indicated by adding the letter m (printed in Roman type) to the mass number of the nuclide.

Examples:

1. ^{235}U   2. ^{3}H   3. ^{14}C   4. ^{133m}Xe

The atomic symbols used in the nuclide symbols are those given in the reference in Note 2d. In the nuclide symbol, the atomic symbol is also printed in Roman type. For the hydrogen isotopes, protium, deuterium and tritium, the nuclide symbols ^{1}H, ^{2}H, and ^{3}H, respectively, are used. The symbols D and T for ^{2}H and ^{3}H, respectively, may be used, but not when other labelling nuclides are also present because this may cause difficulties in the alphabetical ordering of the nuclide symbols in the isotopic descriptor.

II-2.2.4 Names for hydrogen atoms, ions and groups

These names comprise general names, to be used without regard to the nuclear mass of the hydrogen entity, either for hydrogen in its natural isotopic abundance or where it is not desired to distinguish between the isotopes, and specific names pertaining to nuclides (Table II-2.1).

Table II-2.1 Names of hydrogen atoms, ions or groups

	General	^{1}H	^{2}H or D	^{3}H or T
Atom (H)	hydrogen	protium	deuterium	tritium
Cation (H^{+}) (Note 2f)	hydron	proton	deuteron	triton
Anion (H^{-})	hydride	protide	deuteride	tritide
Group (–H)	hydrido hydro (in boron compounds	protio	deuterio	tritio

II-2.2.5 Order of nuclide symbols

When it is necessary to cite different nuclides at the same place in the formula or name of an isotopically modified compound, the nuclide symbols are written in alphabetical order according to their atomic symbols; and when their atomic symbols are identical, in order of increasing mass number. The atomic symbols, with their locants, if any, are separated from each other by a comma.

Note 2f. Proton, deuteron and triton are the names of particles in nuclear physics and nuclear chemistry and their symbols are p, d and t, respectively (see Note 2a, p. 43).

II-2.3 ISOTOPICALLY SUBSTITUTED COMPOUNDS

II-2.3.1 Definition of isotopically substituted compounds

An isotopically substituted compound has a composition such that essentially all the molecules of the compound have only the indicated nuclide(s) at each designated position. For all other positions, the absence of nuclide indication means that the nuclide composition is the natural one.

II-2.3.2 Formulae of isotopically substituted compounds

The formula of an isotopically substituted compound is written as usual except that appropriate nuclide symbols are used. When different nuclides of the same element are present at the same position, their symbols should be written in order of increasing mass number (see also Section II-2.2.5).

II-2.3.3 Names of isotopically substituted compounds

The name of an isotopically substituted compound is formed by inserting in parentheses the appropriate nuclide symbol(s), preceded by any necessary locant(s) (letters and/or numerals) before the name or, preferably, before the name for that part of the compound that is isotopically substituted. Immediately after the parentheses there is neither space nor hyphen, except that when the name or a part of the name includes a preceding locant, a hyphen is inserted (see Note 2g).

When isotopic polysubstitution is possible, the number of atoms that have been substituted is always specified as a right subscript to the atomic symbol(s) even in the case of monosubstitution.

When different nuclides must be cited at the same place in the name of an isotopically substituted compound, the nuclide symbols are ordered as prescribed by Section II-2.2.5.

Examples:

Isotopically unmodified compound	*Isotopically modified compound*
1. H_2O	H^3HO
water	($^3H_1$)water
2. Br_2	$^{78}Br^{81}Br$
dibromine	($^{78}Br,^{81}Br$)dibromine
3. H_2SO_4	$^2H_2^{35}SO_4$
sulfuric acid	($^2H_2,^{35}S$)sulfuric acid
4. $NaCl$	$Na^{36}Cl$
sodium chloride	sodium (^{36}Cl)chloride
5. UF_6	$^{235}UF_6$
uranium hexafluoride	(^{235}U)uranium hexafluoride
uranium(VI) fluoride	(^{235}U)uranium(VI) fluoride

Note 2g. All IUPAC recommended names can be used (see Note 2d). In the examples usually only one is mentioned.

6. InCl$_3$ ^{113m}InCl$_3$
   indium trichloride   (^{113m}In)indium trichloride
   indium(III) chloride   (^{113m}In)indium(III) chloride
   indium(3 +) chloride   (^{113m}In)indium(3 +) chloride
7. SiH$_4$ SiH$_3^2$H
   silane   (^{2}H$_1$)silane
8. SiH$_3$–SiHCl–SiH$_3$ SiH$_2^2$H–SiH^{36}Cl–SiH$_3$
               1   2   3

   2-chlorotrisilane   2-[(^{36}Cl)chloro](1-^{2}H$_1$)trisilane
       (for numbering see Section II-2.5.3)
9. N$_2$O ^{15}N$_2$O
   dinitrogen monoxide   di[(^{15}N)nitrogen] monoxide
10. KNaCO$_3$ ^{42}KNa^{14}CO$_3$
    potassium sodium carbonate   (^{42}K)potassium sodium (^{14}C)carbonate
11. Ca$_3$(PO$_4$)$_2$ Ca$_3$(^{32}PO$_4$)$_2$
    tricalcium bis(phosphate)   tricalcium bis[(^{32}P)phosphate]
12. K[PF$_6$]   K[^{32}PF$_6$]
    potassium hexafluorophosphate   potassium hexafluoro(^{32}P)phosphate
13. [Cr(H$_2$O)$_6$]Cl$_3$ [^{50}Cr(^{2}H$_2$O)$_6$]Cl$_3$
    hexaaquachromium trichloride   hexa[(^{2}H$_2$)aqua](^{50}Cr)chromium trichloride

In order to retain the same name for an isotopically modified compound as for the corresponding unmodified compound as far as possible, an isotopic descriptor may be placed before a numerical prefix.

Examples:

14. BCl$_3$ B^{35}Cl^{37}Cl$_2$
    boron trichloride   boron (^{35}Cl$_1$,^{37}Cl$_2$)trichloride
    trichloroborane   (^{35}Cl$_1$,^{37}Cl$_2$)trichloroborane
15. K$_4$[Fe(CN)$_6$]   K$_3^{42}$K[Fe(CN)$_6$]
    tetrapotassium hexacyanoferrate   (^{42}K$_1$)tetrapotassium hexacyanoferrate

II-2.4   ISOTOPICALLY LABELLED COMPOUNDS

II-2.4.1   Definition of isotopically labelled compounds

An isotopically labelled compound may be considered formally as a mixture of an isotopically unmodified compound and one or more analogous isotopically substituted compounds (Note 2h).

Isotopically labelled compounds may be of various types, such as:

(a) specifically labelled (see Section II-2.4.2);
(b) selectively labelled (see Section II-2.4.3);

Note 2h. Although an isotopically labelled compound is really a mixture as far as chemical identity is concerned (in the same way as is an unmodified compound), for nomenclature purposes such a mixture is called an isotopically labelled compound.

(c) non-selectively labelled (see Section II-2.4.4); and
(d) isotopically deficient (see Section II-2.4.5).

II-2.4.2 Specifically labelled compounds

II-2.4.2.1 *Definition of specifically labelled compounds*

An isotopically labelled compound is called specifically labelled when a unique isotopically substituted compound is added formally to the analogous isotopically unmodified compound. In such a case, both position(s) and number of each labelling nuclide are defined (Note 2i).

II-2.4.2.2 *Formulae of specifically labelled compounds*

The formula of a specifically labelled compound is written in the usual way, but with the appropriate nuclide symbol(s) and multiplying subscript, if any, enclosed in square brackets. When the different nuclides of the same element are present at the same place, the nuclide symbols are ordered according to Section II-2.2.5 (Note 2j).

Examples:

Isotopically substituted compound	when added to	isotopically unmodified compound	results in	specifically labelled compound
1. $H^{36}Cl$		HCl		$H[^{36}Cl]$
2. $H^{99m}TcO_4$		$HTcO_4$		$H[^{99m}Tc]O_4$
3. $Ge^2H_2F_2$		GeH_2F_2		$Ge[^2H_2]F_2$
4. $^{32}PCl_3$		PCl_3		$[^{32}P]Cl_3$

II-2.4.2.3 *Names of specifically labelled compounds*

The name of a specifically labelled compound is formed by inserting in square brackets the nuclide symbol(s), preceded by any necessary locant(s) (letters and/or numerals), before the name or preferably before the name for that part of the compound that is isotopically modified. Immediately after the square brackets there is neither space nor hyphen except that when the name, or a part of the name, requires a preceding locant a hyphen is inserted (see Note 2g).

When it is possible to label more than one atom of the same element, the number of atoms that have been labelled is always specified as a right subscript to the atomic symbol(s) even when only one is labelled. This is necessary in order to distinguish between a specifically and a selectively or non-selectively labelled compound.

Note 2i. When free exchange among atoms of the same element in a compound occurs, *e.g.* H in NH_3 and in H_2NNH_2 when in contact with aqueous media, specific labelling is not possible and such isotopically labelled compounds must be considered as selectively or non-selectively labelled (see Sections II-2.4.3 and II-2.4.4).

Note 2j. Although the formula for a specifically labelled compound does not represent the composition of the bulk material, which usually consists overwhelmingly of the isotopically unmodified compound, it does indicate the presence of the compound of chief interest, the isotopically labelled compound.

ISOTOPICALLY MODIFIED INORGANIC COMPOUNDS

When different nuclides must be cited at the same place in the name of a specifically labelled compound, the nuclide symbols are ordered as given by Section II-2.2.5.

The name of a specifically labelled compound differs from that of the corresponding isotopically substituted compound (see Section II-2.3.3) only in the use of square brackets instead of parentheses surrounding the nuclide descriptor.

Examples:

Isotopically unmodified compound	Specifically labelled compound
1. HCl	H[^{36}Cl]
hydrogen chloride	hydrogen [^{36}Cl]chloride
2. HTcO$_4$	H[^{99m}Tc]O$_4$
hydrogen tetraoxotechnetate(1–)	hydrogen tetraoxo[^{99m}Tc]technetate(1–)
3. PCl$_3$	[^{32}P]Cl$_3$
phosphorus trichloride	[^{32}P]phosphorus trichloride
4. Na$_2$S$_2$O$_3$	Na$_2$[^{35}S]O$_3$S
sodium thiosulfate	sodium thio[^{35}S]sulfate
5. Na$_2$S$_2$O$_3$	Na$_2$SO$_3$[^{35}S]
sodium thiosulfate	sodium ([^{35}S]thio)sulfate
6. NH$_3$	[^{15}N]H$_2$[^{2}H]
ammonia	[^{2}H$_1$,^{15}N]ammonia
7. GeH$_2$F$_2$	Ge[^{2}H$_2$]F$_2$
difluorogermane	difluoro[^{2}H$_2$]germane

II-2.4.2.4   *Types of specifically labelled compounds*

A specifically labelled compound is:
(a) Singly labelled when the isotopically substituted compound has only one isotopically modified atom.

Examples:

Isotopically unmodified compound	Specifically labelled compound
1. H$_2$O$_2$	HO[^{18}O]H
hydrogen peroxide	hydrogen [^{18}O$_1$]peroxide
2. H$_2$NNH$_2$	H$_2$[^{15}N]NH$_2$
hydrazine	[^{15}N$_1$]hydrazine
3. [Cr(NH$_3$)$_6$]Cl$_3$	[[^{55}Cr](NH$_3$)$_6$]Cl$_3$
hexaamminechromium trichloride	hexaammine[^{55}Cr]chromium trichloride

(b) Multiply labelled when the isotopically substituted compound has more than one modified atom of the same element at the same position, or at different positions.

Examples:

4. SiH$_3$–SiH$_3$	SiH[^{2}H$_2$]–SiH$_3$
disilane	[1,1-^{2}H$_2$]disilane
5. [FeBr$_2$(CO)$_4$]	[FeBr$_2$(CO)$_2$([^{13}C]O)$_2$]
dibromotetracarbonyliron	dibromo([^{13}C$_2$]tetracarbonyl)iron

(c) Mixed labelled when the isotopically substituted compound has more than one kind of modified atom.

Examples:
6. CO_2 $[^{13}C]O[^{17}O]$
carbon dioxide $[^{13}C]$carbon $[^{17}O_1]$dioxide
7. POF_3 $[^{32}P]O[^{18}F_3]$
phosphoryl trifluoride $[^{32}P]$phosphoryl $[^{18}F_3]$trifluoride

II-2.4.3 Selectively labelled compounds

II-2.4.3.1 *Definition of selectively labelled compounds*

An isotopically labelled compound is called selectively labelled when a mixture of isotopically substituted compounds is formally added to the analogous isotopically unmodified compound in such a way that the position(s), but not necessarily the number, of each labelling nuclide is defined. A selectively labelled compound may be considered as a mixture of specifically labelled compounds.

A selectively labelled compound may be:

(a) Multiply labelled when isotopic modification occurs at more than one atom, either in a set of equivalent atoms at one site, *e.g.* H in SiH_4, or at different sites in a molecule, *e.g.* B in $[B_6H_6]^{2-}$;

(b) Mixed labelled when there is more than one labelling nuclide in the compound, *e.g.* B and C in $B_8C_2H_{10}$ (Note 2k).

II-2.4.3.2 *Formulae of selectively labelled compounds*

A selectively labelled compound cannot be described by a unique structural formula. Therefore it is represented by inserting the nuclide symbol(s) preceded by any necessary locant(s) (letters and/or numerals) but without multiplying subscripts, enclosed in square brackets directly before the usual formula, or, if necessary, before parts of the formula that have an independent numbering. Identical locants are not repeated.

When different nuclides are present at the same place in the formula, the nuclide symbols are ordered according to Section II-2.2.5.

Examples:

Mixture of isotopically substituted compounds	when added to	isotopically unmodified compound	results in	selectively labelled compound
1. $SOCl^{36}Cl$ $SO^{36}Cl_2$		$SOCl_2$		$[^{36}Cl]SOCl_2$

Note 2k. When there is only one atom of an element in a compound that can be modified, only specific labelling can result (see Section II-2.4.2).

2. $H_3^{32}PO_3^{18}O$
   $H_3^{32}PO_4$
   $H_3PO_3^{18}O$ H_3PO_4 $[^{18}O,^{32}P]H_3PO_4$
   $H_3PO_2^{18}O_2$
   $H_3PO^{18}O_3$
   etc., or any two of the above

3. PH_2^2H, PH^2H_2, P^2H_3 PH_3 $[^2H]PH_3$
   or any two of the above

4. $H_2^2HSi\text{–}SiH_3$
   $H^2H_2Si\text{–}SiH_3$ $H_3Si\text{–}SiH_3$ $[1\text{-}^2H]Si_2H_6$
   $^2H_3Si\text{–}SiH_3$
   or any two of the above

The method of writing formulae as given by the above recommendation may also be of use if a compound is represented by its molecular formula rather than its structural formula.

II-2.4.3.3   *Names of selectively labelled compounds*

The name of a selectively labelled compound is formed in the same way as the name of a specifically labelled compound (see Section II-2.4.2.3) except that the multiplying subscripts following the atomic symbols are generally omitted except as described in Section II-2.4.3.4. Identical locants corresponding to the same element are not repeated.

The name of a selectively labelled compound differs from the name of the corresponding isotopically substituted compound in the use of square brackets surrounding the nuclide descriptor rather than parentheses and in the omission of repeated identical locants and multiplying subscripts.

Examples:
1. $[^{36}Cl]SOCl_2$ $[^{36}Cl]$sulfinyl chloride
2. $[^{18}O,^{32}P]H_3PO_4$ $[^{18}O,^{32}P]$phosphoric acid
3. $[^2H]PH_3$ $[^2H]$phosphane or $[^2H]$phosphine
4. $[1\text{-}^2H]Si_2H_6$ $[1\text{-}^2H]$disilane
5. $[^{13}C][Fe(CO)_5]$ $[^{13}C]$pentacarbonyliron
6. $[^{15}N]K_3[Fe(CN)_6]$ potassium $[^{15}N]$hexacyanoferrate(III)

II-2.4.3.4   *Indication of the number of labelling nuclides*

In a selectively labelled compound formally arising from mixing an isotopically unmodified compound with several known isotopically substituted compounds, the number or the possible number of labelling nuclide(s) for each position may be indicated by subscripts to the atomic symbol(s) in the isotopic descriptor in both the formula and name. Two or more subscripts referring to the same nuclide symbol are separated by a semicolon. For a multiply or mixed labelled compound (see Section II-2.4.3.1), the subscripts are written successively in the same order as the various isotopically substituted compounds are considered. The subscript zero is used to indicate that one of the isotopically substituted compounds is not modified at the indicated position.

ISOTOPICALLY MODIFIED INORGANIC COMPOUNDS

Examples (for numbering see Section II-2.5.3):

A known mixture of isotopically substituted compounds	when added to	isotopically unmodified compound	results in	selectively labelled compound
1. $SiH_2{}^2HOSiH_2OSiH_3$ $SiH^2H_2OSiH_2OSiH_3$		$H_3SiOSiH_2OSiH_3$		$[1\text{-}^2H_{1;2}]SiH_3OSiH_2OSiH_3$ $[1\text{-}^2H_{1;2}]$trisiloxane
2. $SiH^2H_2OSiH_2OSiH_3$ $SiH^2H_2{}^{18}OSiH_2OSiH_3$				$[1,1\text{-}^2H_{2;2},2\text{-}^{18}O_{0;1}]SiH_3OSiH_2OSiH_3$ $[1,1\text{-}^2H_{2;2},2\text{-}^{18}O_{0;1}]$trisiloxane
3. $SiH_3{}^{18}OSiH_2OSiH_3$ $SiH^2H_2OSiH_2OSiH_3$				$[1\text{-}^2H_{0;2},2\text{-}^{18}O_{1;0}]SiH_3OSiH_2OSiH_3$ $[1\text{-}^2H_{0;2},2\text{-}^{18}O_{1;0}]$trisiloxane

II-2.4.3.5 General labelling

In a selectively labelled compound where all atoms of a particular element are isotopically modified, but not necessarily uniformly, the italicized descriptor *gen*, to indicate a general labelling, may be added immediately preceding, without a hyphen, the nuclide symbol in the isotopic descriptor of the name or formula (Note 21).

Examples:

1. $[^{13}C][Fe(CO)_5]$ where each carbonyl ligand is labelled with ^{13}C, but not necessarily uniformly, may be designated as:

   $[gen^{13}C][Fe(CO)_5]$ $[gen^{13}C]$pentacarbonyliron

2. $[^{190}Os][Os_6(CO)_{18}]$ where each osmium atom is labelled with ^{190}Os, but not necessarily uniformly, may be designated as:

   $[gen^{190}Os][Os_6(CO)_{18}]$ $[gen^{190}Os]$octadecacarbonylhexaosmium

II-2.4.3.6 Uniform labelling

In a selectively labelled compound where all atoms of a particular element are labelled in the same isotopic ratio, the italicized descriptor *unf*, to indicate a uniform labelling, may be added immediately preceding, without a hyphen, the nuclide symbol in the isotopic descriptor of the name or formula (see Note 21).

Examples:

1. $[^{13}C][Fe(CO)_5]$ where the ^{13}C is equally distributed in each carbonyl ligand may be designated as:

   $[unf^{13}C][Fe(CO)_5]$ $[unf^{13}C]$pentacarbonyliron

2. $[^{190}Os][Os_6(CO)_{18}]$ where the ^{190}Os is equally distributed among the six osmium atoms may be designated as:

   $[unf^{190}Os][Os_6(CO)_{18}]$ $[unf^{190}Os]$octadecacarbonylhexaosmium

The italicized descriptor *unf* may be followed by appropriate locants to indicate uniform labelling at specified positions.

Note 21. The recommendations for nomenclature of isotopically modified organic compounds (see Note 2c) use the symbol C and U for general and uniform labelling, respectively, rather than *gen* and *unf* recommended here.

Example:

3. $[^{32}\text{Si}]\overset{1}{\text{Si}}\text{H}_2\overset{2}{\text{Cl}}\overset{3}{\text{O}}\text{Si}\text{H}_2\overset{4}{\text{O}}\overset{5}{\text{Si}}\text{H}_3$ where the ^{32}Si is equally distributed between only the terminal silicon atoms may be designated as:

   $[\textit{unf}\text{-}1,5\text{-}^{32}\text{Si}]\text{SiH}_2\text{ClOSiH}_2\text{OSiH}_3$ 1-chloro[*unf*-1,5-^{32}Si]trisiloxane

II-2.4.4 Non-selectively labelled compounds

II-2.4.4.1 *Definition of non-selectively labelled compounds*

An isotopically labelled compound is called non-selectively labelled when both the position(s) and the number of labelling nuclide(s) are undefined. Non-selective labelling is indicated in the formula and name by inserting the nuclide symbol, enclosed in square brackets, directly before the usual formula or name. No preceding locants or subscripts are used (see Note 2m).

Examples:

1. $[^{10}\text{B}]\text{H}_2\text{BH}_2\text{BHCl}$ 1-chloro[^{10}B]diborane(6)
2. $[^{30}\text{Si}]\text{SiH}_3\text{SiH}_2\text{SiH}_3$ [^{30}Si]trisilane

II-2.4.4.2 *Isotopically labelled non-molecular materials*

Isotopically labelled non-molecular materials, such as ionic solids and polymeric substances, where labelling nuclides may be dispersed throughout a crystal lattice or polymer network, are considered non-selectively labelled. This is designated in the formula and name according to Section II-2.4.4.1, but the formula and name should be enclosed in parentheses followed by subscript x.

Examples:

1. $[^{35}\text{Cl}](\text{NaCl})_x$ $[^{35}\text{Cl}]$(sodium chloride)$_x$
2. $[^{235}\text{U}](\text{UO}_2)_x$ $[^{235}\text{U}]$(uranium dioxide)$_x$
3. $[^{29}\text{Si}](\text{SiO}_2)_x$ $[^{29}\text{Si}]$(silicon dioxide)$_x$

II-2.4.5 Isotopically deficient compounds

An isotopically labelled compound is called isotopically deficient when the isotopic content of one or more elements has been depleted, *i.e.* a nuclide is present in less than the natural ratio. An isotopically deficient compound is designated in the formula and name by adding the italicized *def* immediately preceding, without a hyphen, the appropriate nuclide symbol (Note 2n).

Examples:

1. $[\textit{def}\,^{10}\text{B}]\text{H}_3\text{BO}_3$ [*def* ^{10}B]boric acid
2. $[\textit{def}\,^{235}\text{U}]\text{UF}_6$ [*def* ^{235}U]uranium hexafluoride

Note 2m. When the only atoms of an element that can be modified are at the same position in a compound, only specifically or selectively labelling can occur.

Note 2n. Commercial products are available in which one or more isotopes of an element, particularly the elements of lithium, boron, carbon, nitrogen, uranium, and the noble gases, have been depleted. These materials do not contain the natural ratio of isotopic composition and, therefore, if used for scientific research, should be so indicated. In addition, certain other naturally occurring materials, such as meteorites, contain elements deficient in certain nuclides when compared to what is usually considered to be the natural isotopic ratio.

ISOTOPICALLY MODIFIED INORGANIC COMPOUNDS

II-2.5 LOCANTS AND NUMBERING OF ISOTOPICALLY MODIFIED COMPOUNDS

II-2.5.1 Indication of the position of isotopic modifications

Positions of isotopic modifications in an isotopically modified compound are indicated, as far as possible, by the locants normally used for numbering of chains, rings, or clusters of atoms in the corresponding unmodified compound (Notes 2c, 2d). The assignment of locants in an isotopically modified compound should not be changed from that of the corresponding isotopically modified compound.

Examples:

1. $SiF_3-Si^2H-SiF_3$ with Si^2H_3 substituent at position 2
   (positions 1, 2, 3)

   1,1,1,3,3,3-hexafluoro-2-[($^2H_3$)silyl]-(2-$^2H_1$)trisilane
   (unmodified name: 1,1,1,3,3,3-hexafluoro-2-silyltrisilane)

2. ClHB⟨H⟩B[$^2H_2$] (diborane structure, positions 1, 2)

   1-chloro[2,2-$^2H_2$]diborane(6)
   [unmodified name: 1-chlorodiborane(6)]

II-2.5.2 Priority

When there is a choice between longest chains or a choice between equivalent rings in an isotopically unmodified compound, the preferred chain or ring of the corresponding isotopically modified compound is chosen so that the maximum number of modified atoms of groups are included. If a choice still remains, precedence is given to the chain or ring that contains first a nuclide of higher atomic number, and then a nuclide of higher mass number.

Examples (see also Note 2c, recommendation H-3.21):

1. $SiF_3-SiH-Si[^2H]F_2$ with $SiHF_2$ substituent at position 2
   (positions 1, 2, 3)

   2-(difluorosilyl)-1,1,1,3,3-pentafluoro[3-$^2H_1$]trisilane
   [unmodified name: 2-(difluorosilyl)-1,1,1,3,3-
   -pentafluorotrisilane] (see Note 1b, p. 10)

2. $SiH_2Cl-[^{18}O]-SiH-[^{18}O]-SiH_3$ with $O-SiH_2[^2H]$ substituent at position 3
   (positions 1, 2, 3, 4, 5)

   1-chloro-3-([$^2H_1$]siloxy)[2,4-$^{18}O_2$]trisiloxane
   (unmodified name: 1-chloro-3-siloxytrisiloxane)

II-2.5.3 Numbering

When there is a choice between equivalent numberings in an isotopically unmodified compound, the starting point and direction of numbering are chosen for the analogous isotopically modified compound so as to give the lowest locants to the modified atoms or groups considered together as one series in ascending numerical order (see Note 2c, Section C-15.11) without regard to type of nuclide or mass number. If a choice still remains, preference for lowest locants is given first to a nuclide of higher atomic number, and then to a nuclide of higher mass number.

Examples:

1. $\overset{1}{\text{N}}[^2\text{H}_2]–[^{15}\overset{2}{\text{N}}]\text{H}_2$ [1,1-^{2}H$_2$,2-^{15}N]hydrazine

2. $\text{H}_3\overset{1}{\text{Si}}–[^{29}\overset{2}{\text{Si}}]\text{H}_2–\overset{3}{\text{SiH}}[^2\text{H}]–\overset{4}{\text{SiH}}_3$ [3-^{2}H$_1$,2-^{29}Si]tetrasilane

II-2.5.4 Group symbols and prefixes

When isotopic modification occurs in a structure at a position that is not normally assigned a locant, group symbols or italicized prefixes may be used to denote its position.

Examples:

1. HOSO$_2$[^{35}S]H   [^{35}SH]thiosulfuric acid

2. HO$_3$S[^{18}O–^{18}O]SO$_3$H   hydrogen μ-[^{18}O–^{18}O]peroxohexaoxodisulfate(2–)

II-2.5.5 Italicized nuclide symbols

Italicized nuclide symbols and/or italicized atomic symbols may be used as locants to distinguish different nuclides of the same element.

Examples:

1.
$$\text{CH}_3\text{O}-\underset{\underset{\text{OCH}_3}{|}}{\overset{\overset{^{18}\text{O}}{||}}{\text{P}}}-\text{OCH}_3$$

     O,O,O-trimethyl (^{18}O$_1$)phosphate

2. CH$_3$–[^{18}O]–S(O)$_2$–S–CH$_3$ ^{18}*O,S*-dimethyl[^{18}O]thiosulfate

3. [Ru(NH$_3$)$_5$(^{14}N^{15}N)]$^{2+}$ pentaammine[(^{14}N,^{15}N)dinitrogen-*N*]ruthenium(2+)

II-2.6 SUMMARY OF TYPES OF ISOTOPICALLY MODIFIED COMPOUNDS

Type of isotopic modification	Examples, formula and name	Explanatory remarks
Substituted	SiH$_3$^{2}H (^{2}H$_1$)silane	All molecules contain one, and only one, atom of ^{2}H.
Specifically labelled	SiH$_3$[^{2}H] [^{2}H$_1$]silane	Total ^{2}H content greater than natural amount; excess ^{2}H is in singly substituted molecules.
Selectively labelled	[1-^{2}H]Si$_2$H$_6$ [1-^{2}H]disilane	Total ^{2}H content greater than natural amount; excess ^{2}H occurs in two or more substituted molecules and may occur in any number at the specified position in a given molecule.
	[^{2}H$_{1;3}$]SiH$_4$ [^{2}H$_{1;3}$]silane	Total ^{2}H content greater than natural amount; excess ^{2}H occurs in two substituted molecules, one with one ^{2}H atom and one with three ^{2}H atoms.
Non-selectively labelled	[^{2}H]Si$_3$H$_8$ [^{2}H]trisilane	Total ^{2}H content greater than natural amount; excess ^{2}H may occur in any number and at any position in one or more substituted molecules.
Isotopically deficient	[*def*^{29}Si]SiHF$_3$ [*def*^{29}Si]trifluorosilane	Total ^{29}Si content is less than the natural amount.

II-3 Metal Complexes of Tetrapyrroles

CONTENTS

II-3.1 Introduction
II-3.2 The construction of systematic names
  II-3.2.1 General description of common metal complexes of tetrapyrroles
  II-3.2.2 The numbering of atoms in the tetradentate tetrapyrrole ligand
  II-3.2.3 Coordination nomenclature
  II-3.2.4 Stereochemical descriptors
II-3.3 Trivial names
II-3.4 Less common structural types
  II-3.4.1 Naming complexes with unknown structures
  II-3.4.2 Uncommon coordination modes
II-3.5 Trivial names of porphyrins, chlorins, chlorophylls, bilanes, fundamental rings and related species
  II-3.5.1 Trivially named porphyrins
  II-3.5.2 Trivially named chlorins
  II-3.5.3 Trivially named isobacteriochlorins
  II-3.5.4 Trivially named chlorophylls
  II-3.5.5 Trivially named bilanes
  II-3.5.6 Trivially named fundamental rings and related structures

II.3.1 INTRODUCTION

The provisional recommendations on the nomenclature of tetrapyrroles were initially published by the IUPAC-IUB Joint Commission on Biochemical Nomenclature (JCBN) in 1979 (Note 3a), and then modified and extended by a later IUPAC-IUB joint publication in 1986 (Note 3b). While a systematic approach for naming porphyrins and their derivatives has been developed in order to reduce the number of trivial names still in the current literature, those names based on the Fischer system (Note 3c) are so well established that their use is likely to continue for some time. Such names are therefore retained for the present; some of the more important are listed in Section II-3.5.

Note 3a. Nomenclature of Tetrapyrroles, Recommendations 1978, *Pure Appl. Chem.*, **51**, 2251 (1979); *Eur. J. Biochem.* **108**, 1 (1980).

Note 3b. Nomenclature of Tetrapyrroles, Recommendations 1986, *Pure Appl. Chem.*, **59**, 779 (1987); *Eur. J. Biochem.* **178**, 277 (1988).

Note 3c. H. Fischer and H. Orth, *Die Chemie des Pyrrols*, Volume II.1, Akademische Verlagsgesellschaft, Leipzig, 1937; H. Fischer and A. Stern, *Die Chemie des Pyrrols*, Volume II.2, Akademische Verlagsgesellschaft, Leipzig, 1940.

II-3.2 THE CONSTRUCTION OF SYSTEMATIC NAMES

II-3.2.1 General description of common metal coordination complexes of tetrapyrroles

A roughly planar tetrapyrrole acts commonly as a tetradentate dianionic ligand and coordinates with a metal ion to form an approximately square planar complex. In several complexes there are additionally two axial monodentate ligands, on each side of the equatorial plane consisting of the metal ion and the tetrapyrrole ligand.

Example:
1.

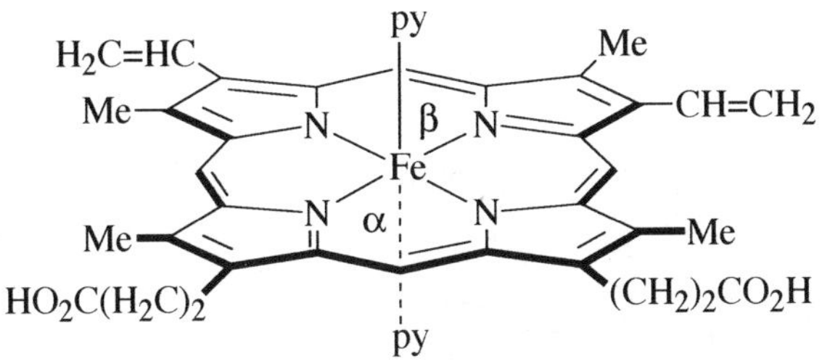

A tetrapyrrole coordination complex containing a tetrapyrrole tetradentate dianionic ligand and two axial pyridine ligands (Me = methyl, py = pyridine).

II-3.2.2 The numbering of atoms in the tetradentate tetrapyrrole ligand

The fundamental macrocyclic tetrapyrrolic ring system has the trivial name porphyrin. This structural unit, together with the 1–24 numbering scheme of the atoms, is shown below. For a detailed discussion on the numbering of tetrapyrroles see Notes 3b and 3d. The basic porphyrin exhibits tautomerism with regard to the locations of the two hydrogen atoms bound to the nitrogen atoms and specific assumptions on their locations need to be made for nomenclature purposes (Note 3b). In the case of the dianionic porphyrinate, the hydrons have been removed and this problem does not exist.

Example:
1.

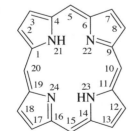

The 1–24 numbering scheme in porphyrins.

Note 3d. R.S. Cahn, C. Ingold and V. Prelog, *Angew. Chem. Int. Ed. Engl.*, **5**, 385 (1966); V. Prelog and G. Helmchen, *Angew. Chem. Int. Ed. Engl.*, **21**, 567 (1983).

II-3.2.3 **Coordination nomenclature**

The basic principles in constructing the coordination name for the complexes follow the guidelines given in Chapter I-10 of Note 3e. The name of the dianionic macrocyclic ligand has the ending -ato. Thus chlorin becomes chlorinato, porphyrin becomes porphyrinato, *etc.* (see Section II-3.5).

The names of the ligands are placed in alphabetical order and precede the name of the central atom. The oxidation state of the metal ion may be indicated by the oxidation number (Stock number), or the charge of the entire coordination entity by the charge number placed in parentheses after the name of the metal.

Examples:
1.

dichloro[2,7,12,17-tetraethyl-3,8,13,18-tetramethylporphyrinato]germanium(IV)

Sometimes it is desirable to designate the atoms coordinated to the metal atom. This can be accomplished by using the kappa (κ) convention (see Section I-10.6.2.2 of Note 3e). In the case of tetrapyrrole metal complexes the superiority of the kappa convention over the earlier practice of suffixing the italicized symbols of the ligating atoms to the name of the ligand is clearly demonstrated. The name of the complex then becomes dichloro(2,7,12,17-tetraethyl--3,8,13,18-tetramethylporphyrinato-$\kappa^4 N^{21,22,23,24}$)germanium(IV) (see Note 1b, p. 10).

The name indicates that the porphyrinato ligand is tetradentate and coordinates through nitrogen atoms, which have the locants 21–24. When the coordination is through all four nitrogens, their locants may be omitted.

2.

The thick black lines denote the planar porphyrin framework.

bis(acetato-κO)(2,3,7,8,12,13,17,18-octaethylporphyrinato-$\kappa^4 N^{21,22,23,24}$)tin(IV)

Note 3e. *Nomenclature of Inorganic Chemistry, Recommendations 1990*, Blackwell Scientific Publications, Oxford, 1990.

3.

chloro(nitrosyl-κN)(5,10,15,20--tetraphenylporphyrinato-κ⁴$N^{21,22,23,24}$)iron

II-3.2.4 Stereochemical descriptors

There are two ways to give stereochemical information about a tetrapyrrole metal complex. The method described in Note 3b is based on viewing the tetrapyrrole ligand in the orientation which enables the clockwise numbering when seen from above. The axial ligand below the tetrapyrrole plane is designated by α and that above the plane by β. The relative positions of the two pyridine ligands are indicated in Example 1 of Section II-3.2.1.

Section I-10.5 of Note 3e presents a general systematic approach for designating stereochemical information in mononuclear coordination entities. It involves the use of the polyhedral symbol (I-10.5.2) together with the configuration index. In the derivation of the configuration index the ligating atoms are assigned priority numbers applying the method of Cahn, Ingold, and Prelog (the so-called CIP rules; see Note 3d and Section I-1.5.3 of Note 3e). The application of the *trans* maximum difference of priority numbers results in the final configuration index used as a prefix in the name.

In the case of tetrapyrrole metal complexes the assignment of priority numbers by CIP rules involves several steps and it is more convenient to use α, β descriptors to convey stereochemical information. However, the procedure presented in Section I-10.5 of Note 3e is generally suitable for all types of mononuclear coordination entities and also provides information about the ideal coordination polydedron.

The assignment of the priority numbers by the CIP rules requires that higher atomic number precedes lower. Thus, oxygen is given the priority number 1.

The five nitrogen atoms in the rings denoted **A–E** above bound to manganese are distinguished by the successive application of the sub-rules in Section I-10.5 of Note 3e. In the following diagrams the connectivities of each nitrogen are inspected in terms of the atomic numbers of the atoms. The priority numbers are assigned as in Note 3d.

Examples:
1.

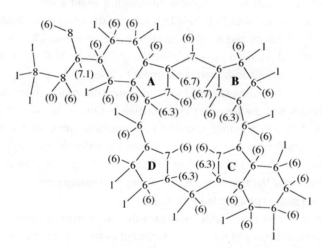

α-(acetato)(phthalocyaninato)-
β-(pyridine)manganese(III)

For the tetrapyrrole ring:

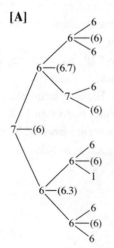

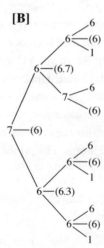

METAL COMPLEXES OF TETRAPYRROLES

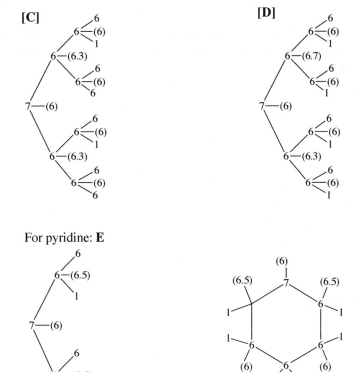

The complete set of priority numbers is thus: O (1); A (2); B (3); E (4); C(5); D(6).

The configuration index in the octahedral complex (*OC*-6) is determined by two digits. The first indicates the priority number of a ligand *trans* to the ligand having the priority number 1. The second digit is the priority number of a ligand *trans* to the ligand with the highest priority number in the plane of four atoms perpendicular to the reference axis defined by the ligand of highest priority (Section I-10.5.4.2 of Note 3e).

The name of the complex is thus (*OC*-6-45)-(acetato)(phthalocyaninato)(pyridine)-manganese(III).

Example:
2.

α,β-bis(pyridine){2,7,12,18-tetramethylporphyrin-3,8,13,17-tetrakis(2-carboxyethyl)}iron(II), or
(*OC*-6–12)bis(pyridine){2,7,12,18-tetramethylporphyrin-3,8,13,17-tetrakis(2-carboxyethyl)}iron(II)

II-3.3 TRIVIAL NAMES

Because of their natural occurrence, the magnesium and iron complexes of tetrapyrroles are associated with an extensive trivial nomenclature. For example there are nine trivial names which may be used to name the chlorophyll structures. A comprehensive list (porphyrins, chlorins, *etc.*) is given in Section II-3.5.

Examples:
1.

	R^1	R^2	R^3
chlorophyll *a*	$R^1 = CH=CH_2$	$R^2 = Me$	$R^3 = $ phytyl
chlorophyll *b*	$R^1 = CH=CH_2$	$R^2 = CHO$	$R^3 = $ phytyl
chlorophyll *d*	$R^1 = CHO$	$R^2 = Me$	$R^3 = $ phytyl
mesochlorophyll α	$R^1 = Et$	$R^2 = Me$	$R^3 = $ phytyl
chlorophyllide α	$R^1 = CH=CH_2$	$R^2 = Me$	$R^3 = H$

Similarly, trivial names for iron complexes may be defined and interrelated as follows:

heme (haem)	iron porphyrin complex
ferroheme	iron(II) porphyrin complex
ferriheme	iron(III) porphyrin complex
hemochrome	low-spin iron porphyrin complex with one or more strong field axial ligands
ferrohemochrome	iron(II) hemochrome
ferrihemochrome	iron(III) hemochrome
hemin	chloro(porphyrinato)iron(III) complex
hematin	hydroxo(porphyrinato)iron(III) complex

2.

protohemin or chloro(protoporphyrinato)iron(III)

METAL COMPLEXES OF TETRAPYRROLES

II-3.4 LESS COMMON STRUCTURAL TYPES

The majority of tetrapyrrole metal complexes involve a square planar coordination of the metal ion to the four nitrogen atoms with one or more axial ligands forming either a square pyramidal or an octahedral coordination polyhedron. There is, however, an increasing number of complexes with differing structural features. They can be named according to the principles described in Chapter I-10 of Note 3e.

II-3.4.1 Naming complexes with unknown structures

Stoichiometric names are used when the structure is unknown. The form 'metal complex of free base' is recommended in the Nomenclature of Tetrapyrroles (Note 3b).

Example:
1. zinc complex of bilirubin   zinc bilirubinate

II-3.4.2 Uncommon coordination modes

If the metal ion is coordinated to an atom other than the four central nitrogen atoms, this must be designated. In the absence of designators the coordination is assumed to be at the four nitrogen atoms.

Examples:

1. tetracarbonyl[(7,8-η)-(21,23-dimethylporphyrin)]iron

2. bis(phthalocyaninato)uranium(iv)

3.

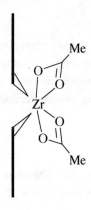

bis(acetato-$\kappa^2 O,O'$)(2,3,7,8,12,13,17,18-octaethylporphyrinato-$\kappa^4 N^{21,22,23,24}$)zirconium(IV)
The thick black lines denote the octaethylporphyrinato ring.

4.

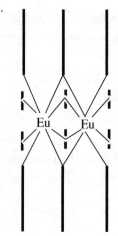

μ-(2,3,7,8,12,13,17,18-octaethylporphyrinato-1$\kappa^4 N^{21,22,23,24}$:2$\kappa^4 N^{21,22,23,24}$)-bis[(2,3,7,8,12,13,17,18-octaethylporphyrinato-$\kappa^4 N^{21,22,23,24}$)europium(III)]
The thick black lines denote the octaethylporphyrinato ring.

5.

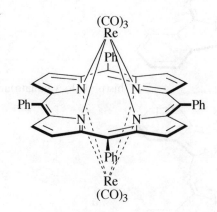

μ-(5,10,15,20-tetraphenylporphyrinato-1$\kappa^4 N^{21,22,23}$:2$\kappa^4 N^{22,23,24}$)bis(tricarbonylrhenium)

The α,β designators can be used to indicate the stereochemistry in polynuclear complexes. The systematic method shown in Section I-10.5 of Note 3e is intended only for mononuclear coordination entities and its elaboration for more extensive structures is yet to be made.

Examples:

1.
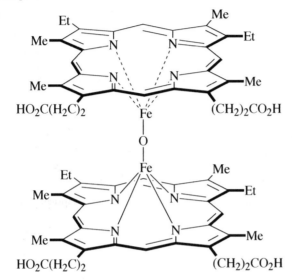

μ-oxo-(α,β)-bis[mesoporphyrinatoiron(III)]

2.

μ-oxo-(α,α)-bis[mesoporphyrinatoiron(III)]

II-3.5 TRIVIAL NAMES OF PORPHYRINS, CHLORINS, CHLOROPHYLLS, BILANES, FUNDAMENTAL RINGS AND RELATED SPECIES (Note 3f)

II-3.5.1 Trivially named porphyrins

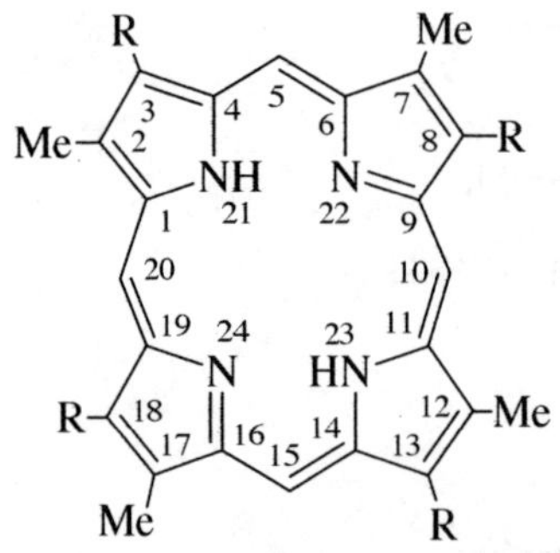

coproporphyrin I (Note 3g)

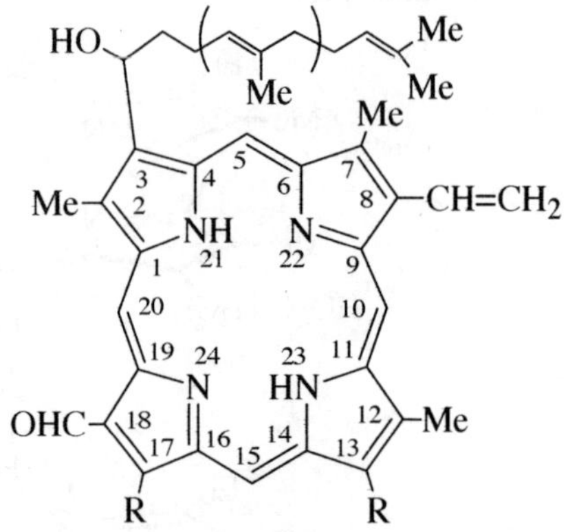

cytoporphyrin

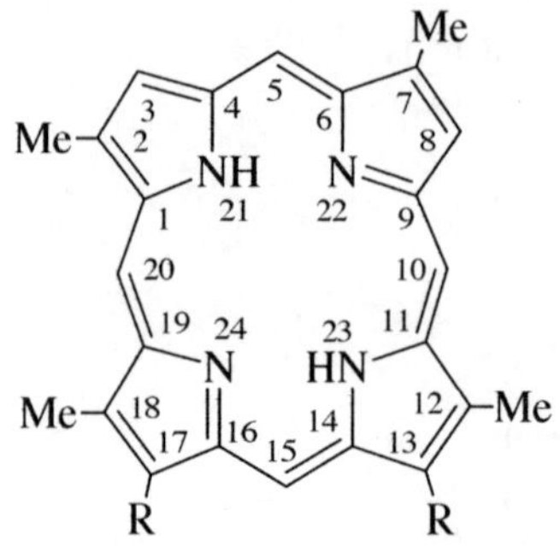

deuteroporphyrin (Note 3h)

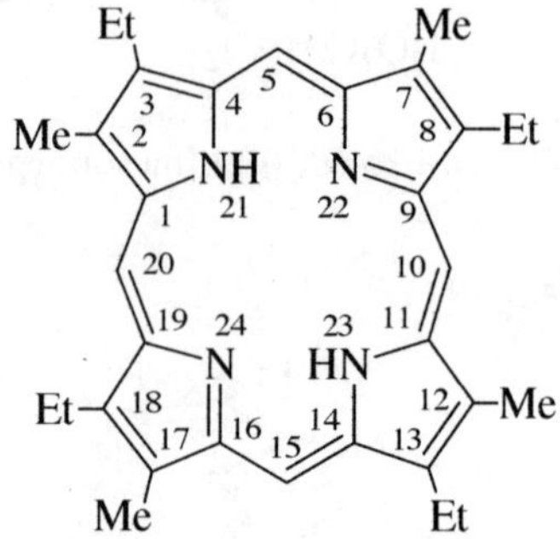

etioporphyrin I (Note 3g)

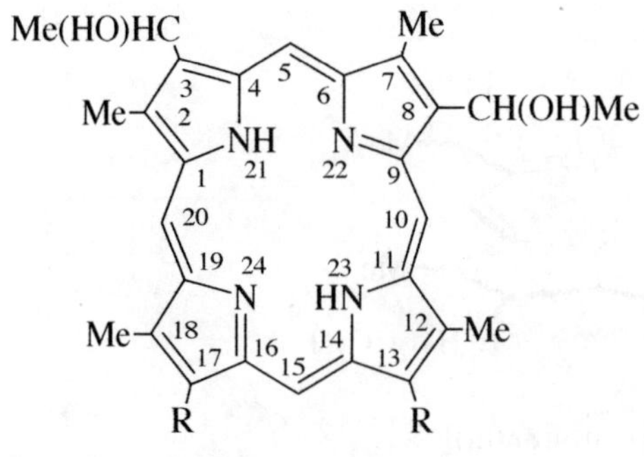

hematoporphyrin

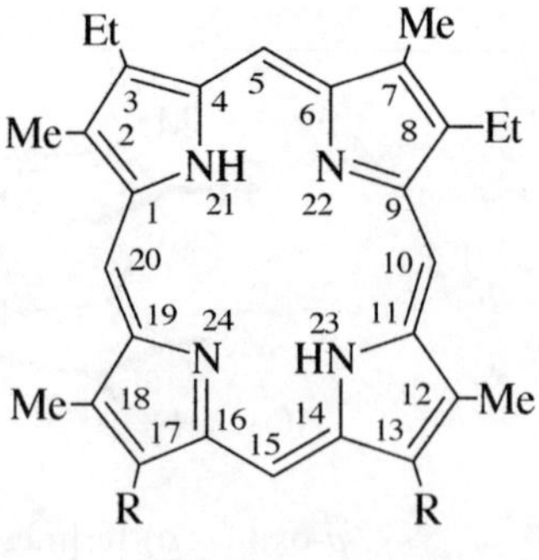

mesoporphyrin

Note 3f. The trivial names apply to the tautomeric structures also (see Note 3a); R = CH$_2$CH$_2$CO$_2$H.
Note 3g. Type I shown only.
Note 3h. Not to be confused with the isotopically labelled compound deuteroporphyrin.

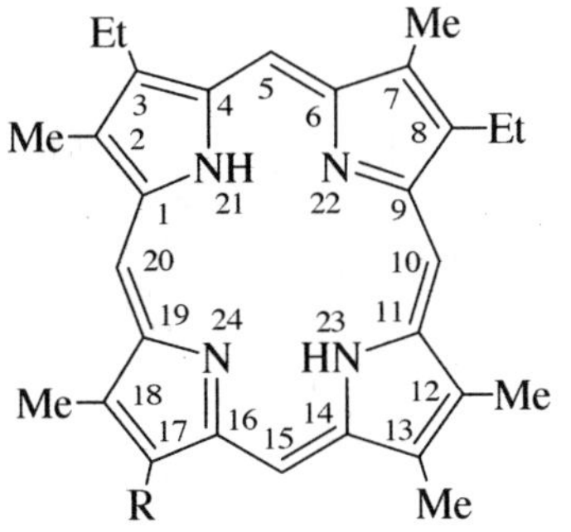

METAL COMPLEXES OF TETRAPYRROLES

phylloporphyrin

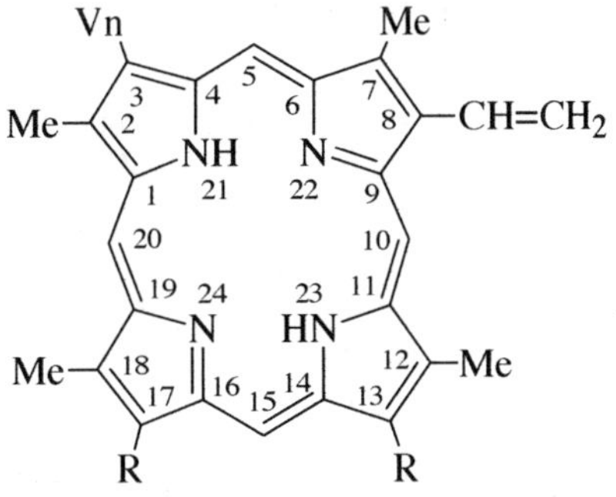

protoporphyrin

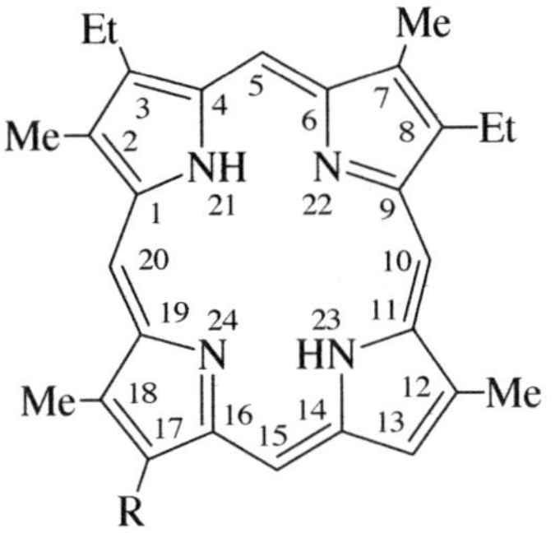

pyrroporphyrin

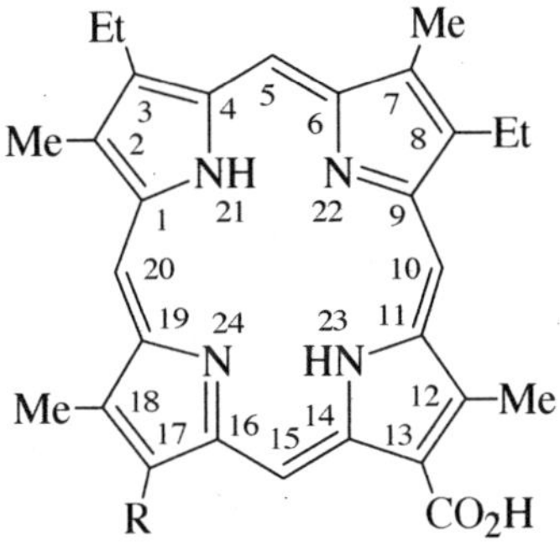

rhodoporphyrin

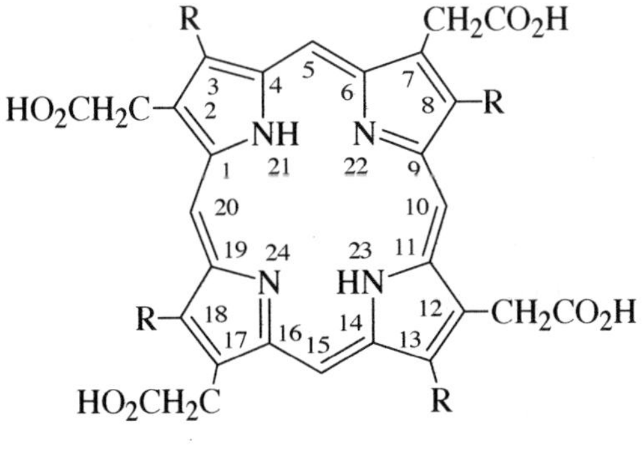

uroporphyrin I (Note 3g)

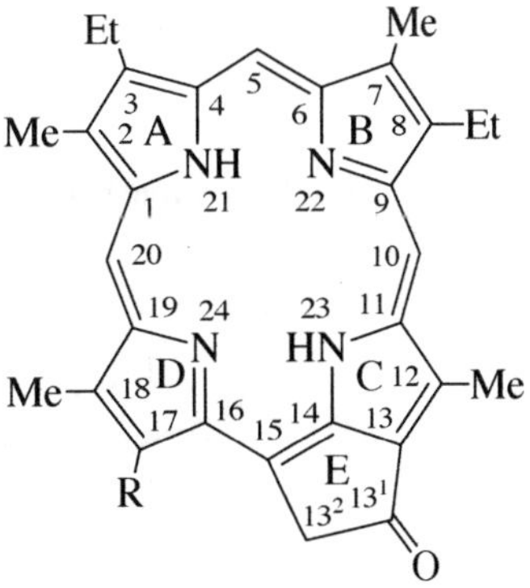

phytoporphyrin

II-3.5.2 Trivially named chlorins

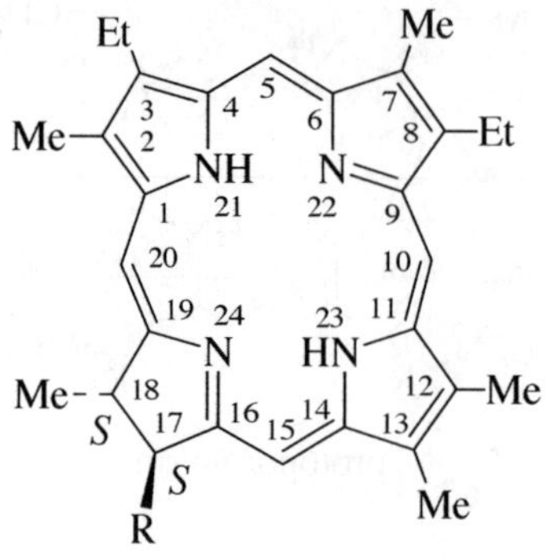

phyllochlorin (Note 3i)

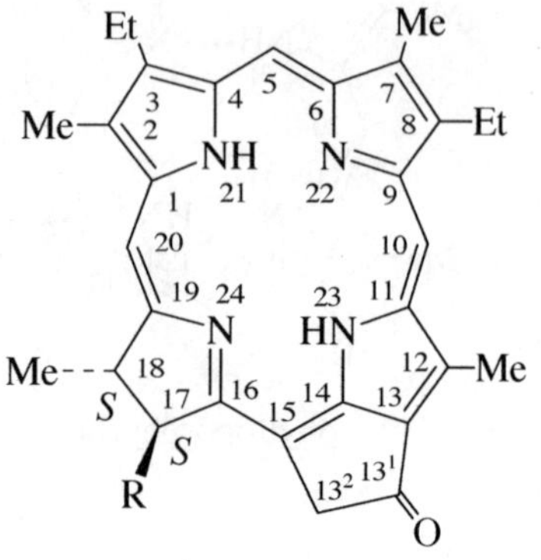

phytochlorin

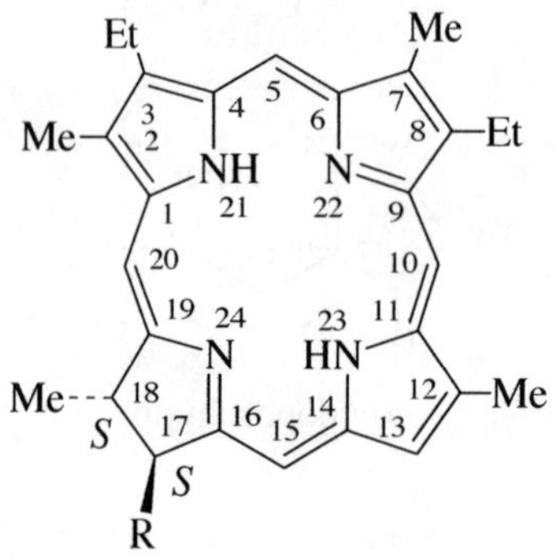

pyrrochlorin

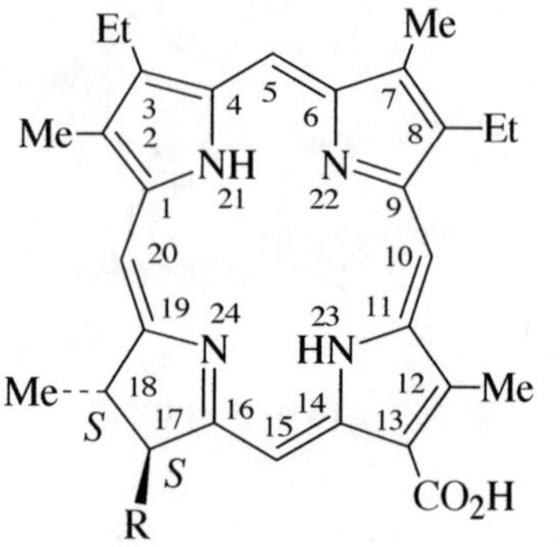

rhodochlorin (Note 3i)

II-3.5.3 Trivially named isobacteriochlorins

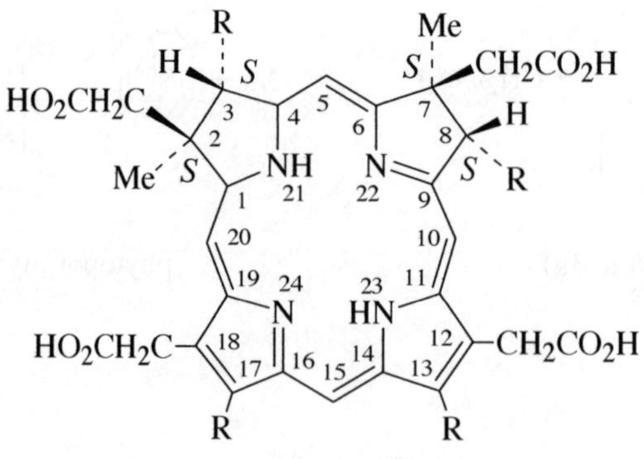

sirohydrochlorin

Note 3i. Phyllochlorin and rhodochlorin have been redefined so that they are now 17,18-dihydro derivatives of the corresponding phylloporphyrin and rhodoporphyrin. These chlorin names have formerly been used in the literature to refer to the structure shown, but with vinyl groups in place of ethyl groups at position 3.

II-3.5.4 Trivially named chlorophylls

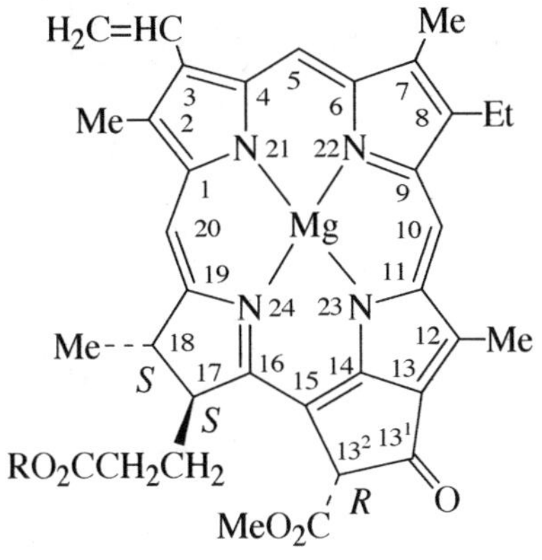

chlorophyll *a* (R = phytyl)

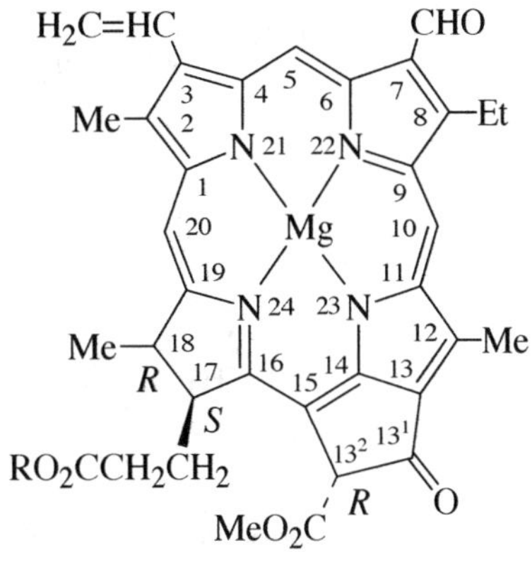

chlorophyll *b* (R = phytyl)

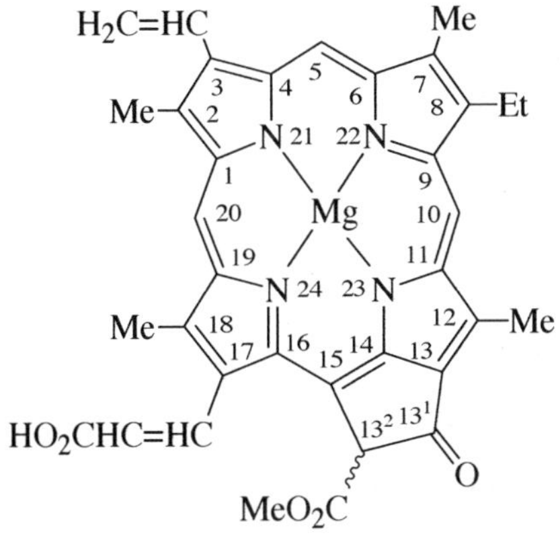

chlorophyll c_1 (Note 3j)

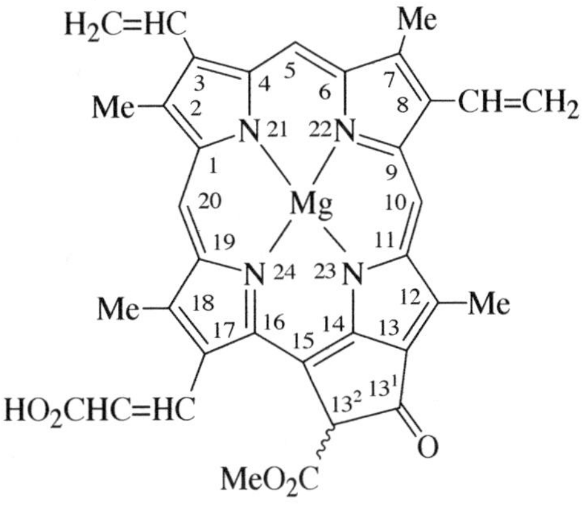

chlorophyll c_2 (Note 3j)

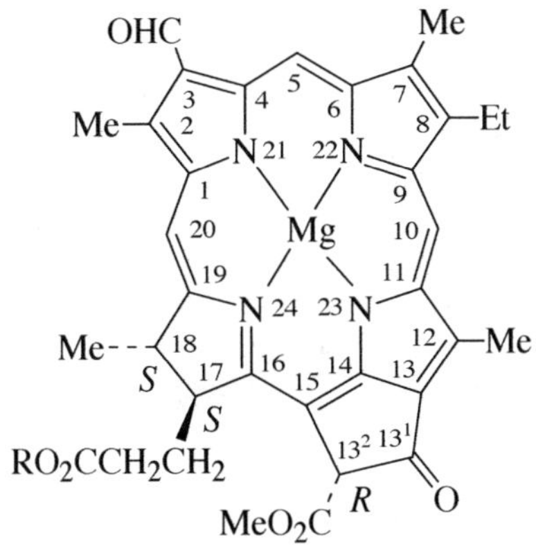

chlorophyll *d* (R = phytyl)

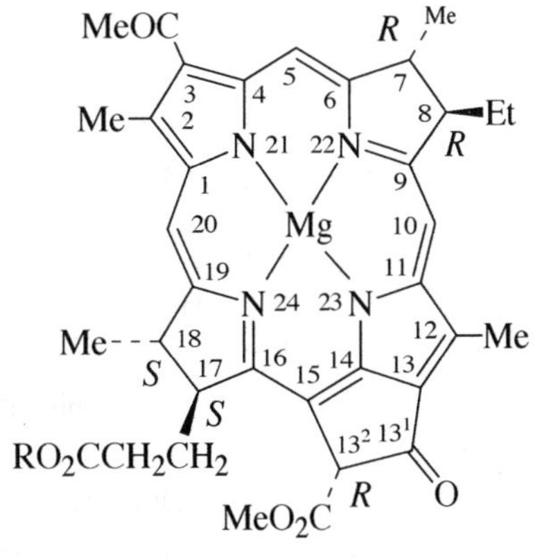

bacteriochlorophyll *a* (R = phytyl)

Note 3j. In contrast to the other chlorophyll structures, chlorophylls c_1 and c_2 are unsaturated at positions 17 and 18. Further, they are free acids at position 17.

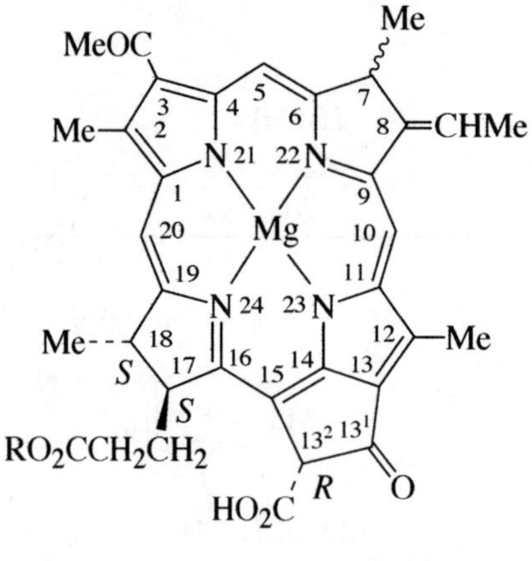

bacteriochlorophyll *b* (R = phytyl)

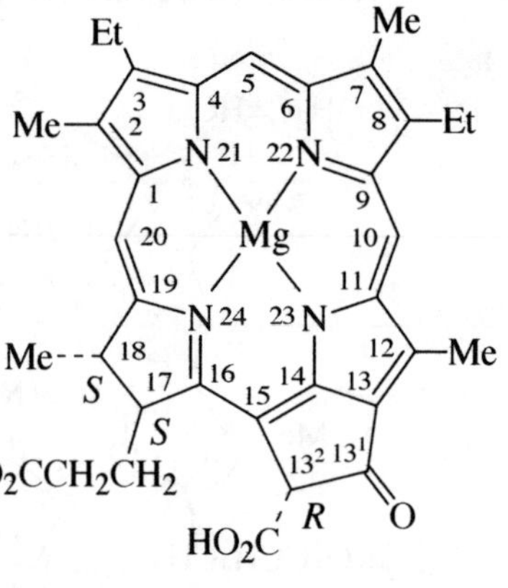

mesochlorophyll *a* (R = phytyl)

II-3.5.5 **Trivially named bilanes**

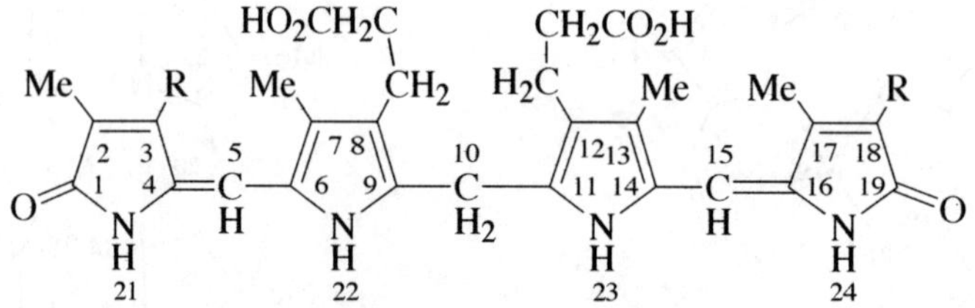

biliverdin (R = vinyl)

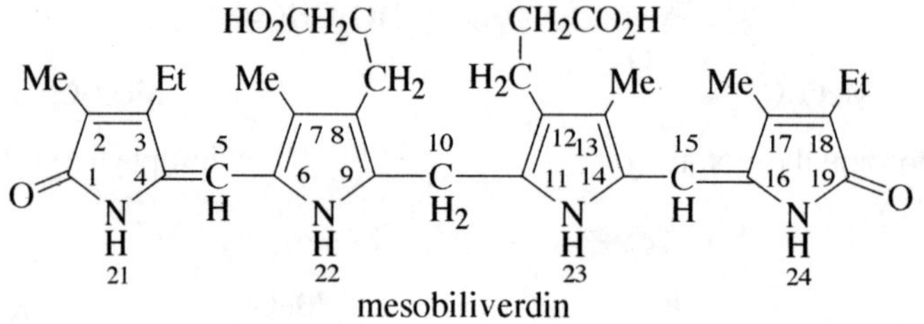

mesobiliverdin

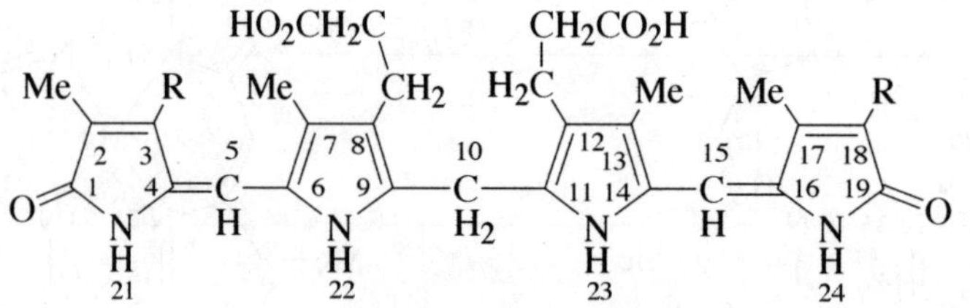

bilirubin (R = vinyl)

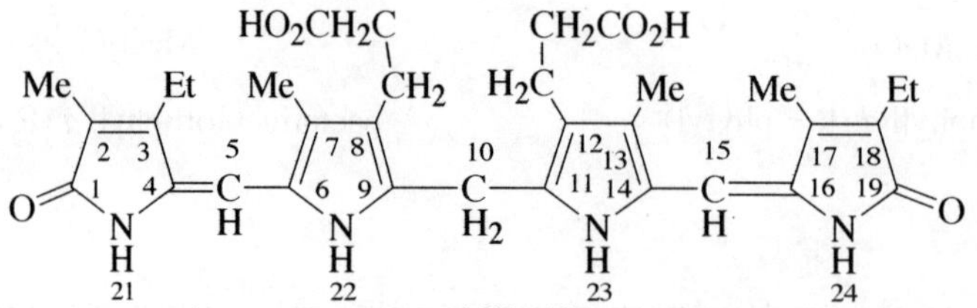

mesobilirubin

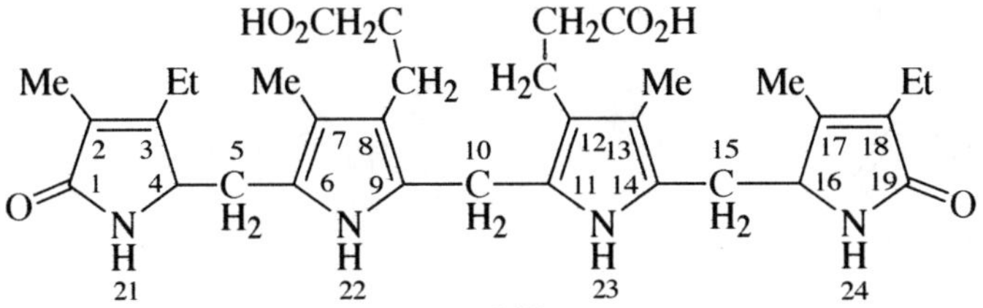

urobilin

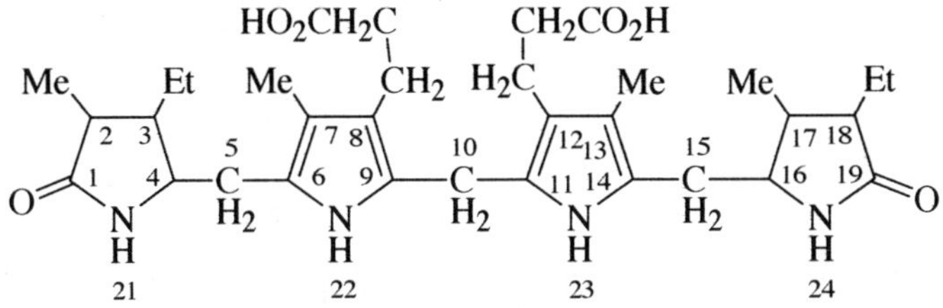

stercobilin

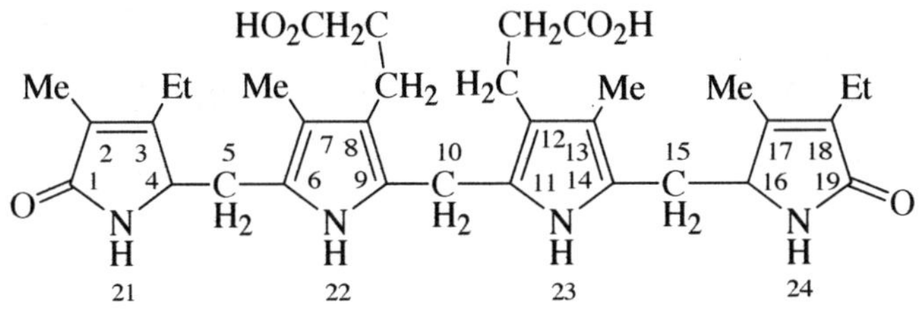

urobilinogen

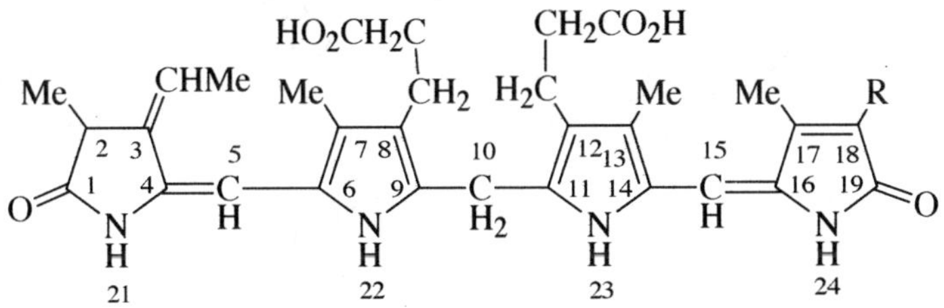

phycoerythrobilin (R = vinyl)

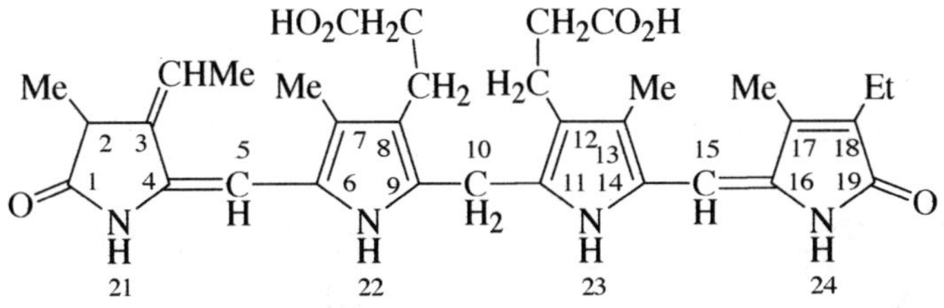

phycocyanobilin

II-3.5.6 Trivially named fundamental rings and related structures

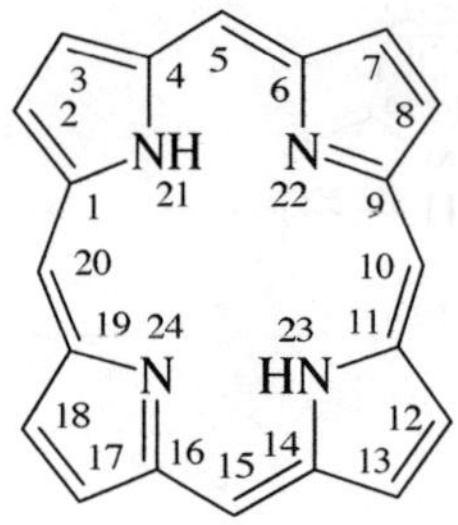

porphyrin

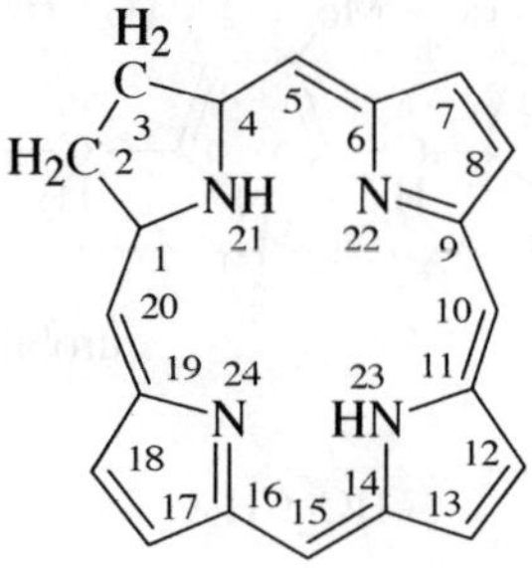

chlorin

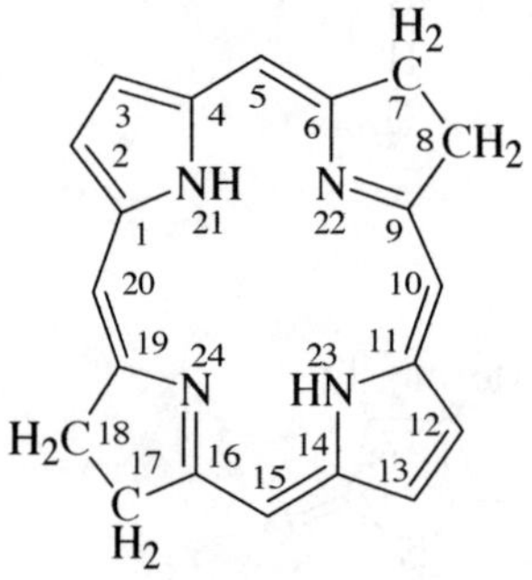

bacteriochlorin

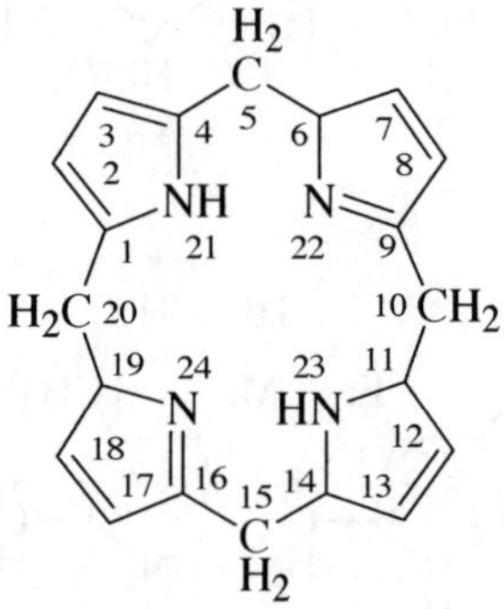

porphyrinogen

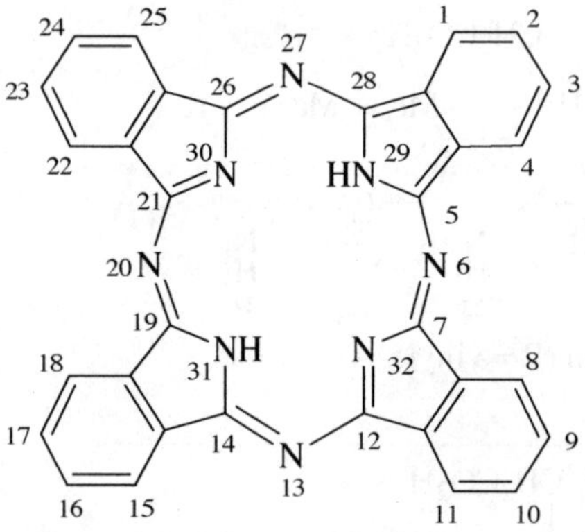

phthalocyanine

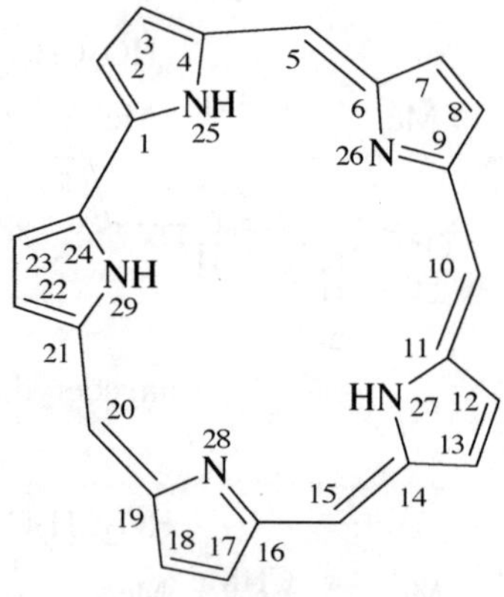

sapphyrin

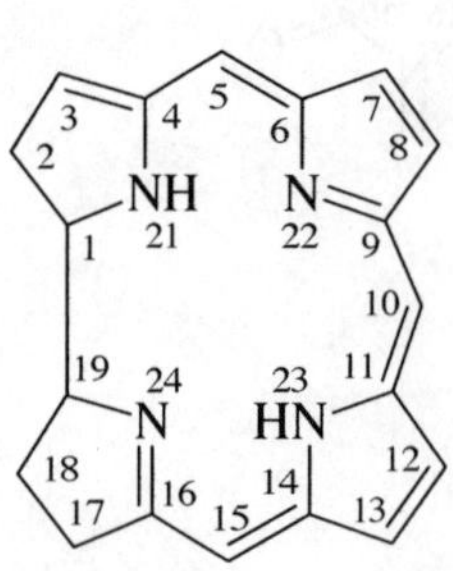

corrin

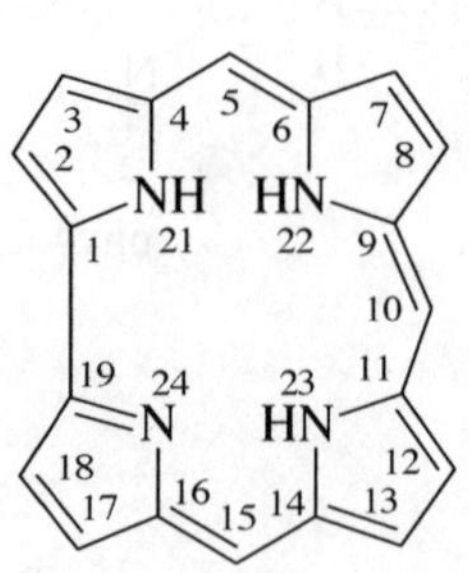

corrole

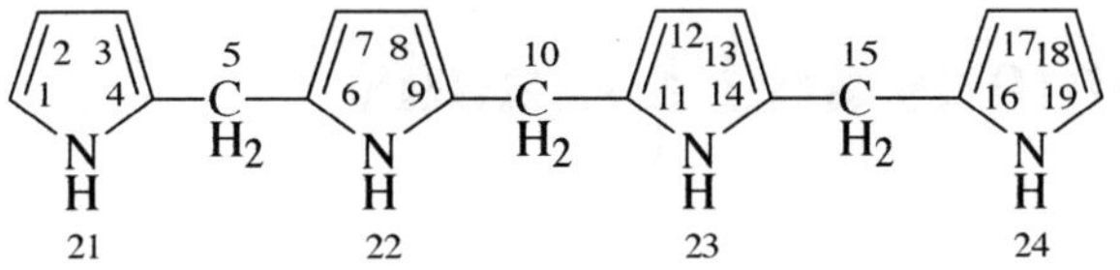

METAL COMPLEXES OF TETRAPYRROLES

bilane

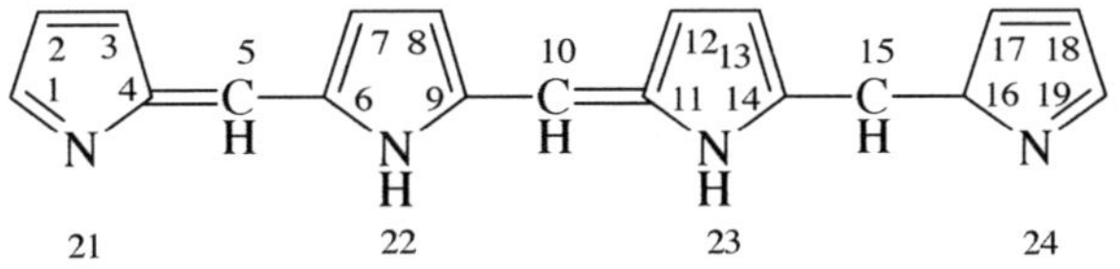

bilin

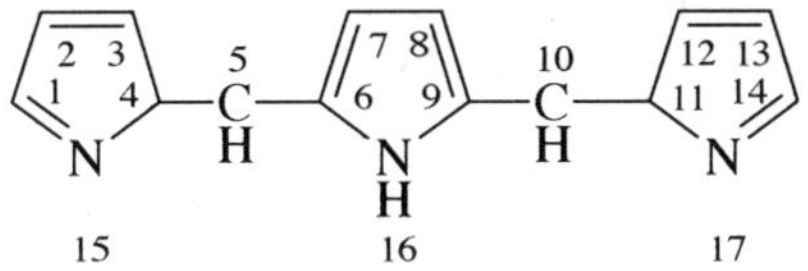

tripyrrin

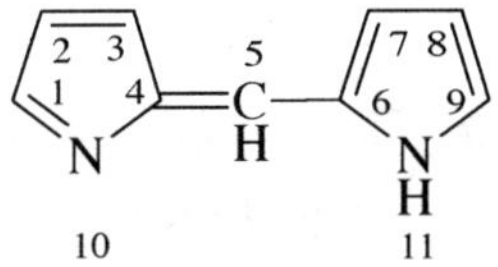

dipyrrin

II-4 Hydrides of Nitrogen and Derived Cations, Anions and Ligands

CONTENTS

II-4.1 Introduction
II-4.2 Parent hydrides
II-4.3 Cations
II-4.4 Anions
II-4.5 Ligands
  II-4.5.1 General
  II-4.5.2 Nitrogen hydrides and their cations as ligands
  II-4.5.3 Ligands formally derived from the nitrogen hydrides by the loss of one or more hydrons
  II-4.5.4 Ligands derived from the nitrogen hydride monocations by loss of one or more hydrons from the uncharged nitrogen centre
II-4.6 Organic derivatives of the nitrogen hydride ligands

II-4.1 INTRODUCTION

The nomenclature of hydrides of nitrogen and derived cations, anions and ligands presents particularly difficult problems. The simple hydrides and many of their derivatives are commercial chemicals with well-established non-systematic traditional and abbreviated names. The hydrides are inorganic but numerous derivatives are organic, and their nomenclature must be compatible with additive (inorganic) and substitutive (organic) nomenclature. It must also be in accord with present trends towards more systematic nomenclature (see Note 4a).

The systematic names are based on organic principles and the hydride name 'azane' where there is hydrogen to be substituted, and on the name 'nitrogen' where there is no hydrogen or only acidic hydrogen.

As the trend towards systematic nomenclature increases these names should eventually take precedence over the well-established names for the common hydrides and their derivatives. For this reason, the trivial names and the non-systematic but well-established derived names are given as alternatives to the systematic names.

II-4.2 PARENT HYDRIDES

Both systematic and common trivial names for nitrogen hydrides are given in Table II-4.1. Trivial names ammonia, hydrazine, and hydrogen azide are still recognized and are exceptions to the systematic hydride nomenclature (Chapter I-7 of Note 4a).

Note 4a. *Nomenclature of Inorganic Chemistry, Recommendations 1990*, Blackwell Scientific Publications, Oxford, 1990.

HYDRIDES OF NITROGEN

Table II-4.1 Names of parent hydrides of nitrogen

Compound	Systematic name	Alternative name
NH_3	azane	ammonia
N_2H_4	diazane	hydrazine
N_2H_2	diazene	diimide
HN_3	hydrogen trinitride	hydrogen azide

II-4.3 CATIONS

Cations derived by the addition of one or more hydrons to a nitrogen or dinitrogen hydride are named by adding -ium to the parent name (see Note 4b) with elision of the final e of the hydride name before i (Section I-8.2.3.3 of Note 4a; Section C-816.2 of Note 4c) except that NH_4^+ is called ammonium. The systematic and trivial names for the most common cations of the parent hydrides are given in Table II-4.2.

If preferred, one unit of charge may be indicated by the use of the charge number (Section I-5.5.2 of Note 4a). Two units of charge *must* be indicated, either by the charge number or by insertion of di- between the hydride name and -ium.

Table II-4.2 The names of the cations derived from the parent hydrides by the addition of one or more hydrons

Cation	Systematic name	Alternative name
NH_4^+	azanium	ammonium
$N_2H_5^+$	diazanium	hydrazinium
$N_2H_6^+$	diazanediium	hydrazinediium
	diazanium(2+)	hydrazinium(2+)
$N_2H_3^+$	diazenium	
$N_2H_4^{2+}$	diazenediium	
	diazenium(2+)	
N_2H^+	diazynium	
$N_2H_2^{2+}$	diazynediium	
	diazynium(2+)	

Examples:
1. NH_4Cl azanium chloride or ammonium chloride
2. $(N_2H_5)_2SO_4$ bis(diazanium) sulfate or hydrazinium sulfate
3. $(N_2H_6)Cl_2$ diazanediium dichloride or hydrazinium(2 +) chloride

Note 4b. The name diazyne is used as a parent name for dinitrogen in organic substitutive nomenclature.
Note 4c. *Nomenclature of Organic Chemistry*, Pergamon Press, Oxford, 1979.

II-4.4 ANIONS

Anions derived by the loss of one or more hydrons from the nitrogen hydrides are named as in Table II-4.3. The charge need not be shown when the anion contains only one nitrogen atom and obeys the octet rule or when it carries only one charge. Otherwise the charge is indicated by the charge number. Alternatively, when the anion contains replaceable hydrogen, the charge can be indicated by the insertion of numerical indicators di-, tri-, *etc.* between the hydride name and -ide.

Table II-4.3 The names of anions derived from the parent hydrides by the loss of one or more hydrons

Anion (Note 4d)	Systematic name	Alternative name
NH_2^-	azanide	amide
NH^{2-}	azanediide azanide(2–)	imide
N^{3-}	nitride	
$N_2H_3^-$	diazanide	hydrazide
$N_2H_2^{2-}$	diazanediide diazanide (2–)	hydrazide(2–)
N_2H^{3-}	diazanetriide diazanide(3–)	hydrazide(3–)
N_2^{4-}	dinitride(4–)	
N_2H^-	diazenide	
N_2^{2-}	dinitride(2–)	
N_3^-	trinitride(1–)	azide

II-4.5 LIGANDS

II-4.5.1 General

There is generally no ambiguity or difficulty in naming ligands derived from the nitrogen hydrides when they occur in main group metal compounds where the bonding situation is well defined. In these compounds conventional ideas regarding valency or formal oxidation states normally lead to an unequivocal choice of names based on structures.

In transition metal compounds the ligand is often intermediate between formal valence bond states, and the formal oxidation state of the metal is not defined. Thus the N_2H_3 ligand may be regarded formally as NH_2NH^- (carrying one negative charge) or $NH_2=NH^+$ (carrying one positive charge). There is no unequivocal possibility and an arbitrary choice must be made for nomenclature purposes. It is important for indexing and information retrieval purposes that the ligand should always have the same name regardless of whether conventional oxidation state ideas dictate otherwise.

Note 4d. When the anion contains no replaceable hydrogen, substitutive names are not needed, and those derived from the name of the element are preferred. However, the substitutive names would be: N^{3-} azanetriide, N_2^{4-} diazanetetraide, N_2^{2-} diazenediide. N_3^- has no approved substitutive name.

HYDRIDES OF NITROGEN

The following order of priorities is used to determine the ligand name:

(i) if possible name the ligand as a neutral molecule which is not a zwitterion, nor a radical, nor a diradical;
(ii) if (i) is impossible, name it as an anionic ligand with the smallest possible formal charge;
(iii) if it cannot be named according to (i) or (ii), name it as a zwitterionic ligand with a total formal charge zero or, failing that, as one with the smallest possible negative charge;
(iv) if none of the above is possible, name the ligand as a cationic ligand.

Some ligands derived from the dinitrogen skeleton have as many as three systematic names depending on the parent chosen for their derivation. Application of the above rules limits the choice to one parent for each N_2H_x ($x = 0$–5) ligand.

II-4.5.2 Nitrogen hydrides and their cations as ligands

The ligand names are the same as those of the hydrides and cations from which they are derived except that ammonia gives the ligand name ammine (Section I-10.4.5.5 of Note 4a).

In earlier practice the points of attachment of the ligand were indicated by suffixing the italicized symbol(s) of the ligating atom or atoms to the name of the ligand (Section I-10.6.2.1 of Note 4a). The kappa, κ, convention (Section I-10.6.2.2 of Note 4a), however, provides a more general system to indicate the points of ligation and, in more complicated cases, may be preferred to the 'donor atom symbol' described above (Note 4e). Ligands in their different bonding situations are given in Table II-4.4.

Table II-4.4 Naming of ligands

Bound ligand	Systematic name	Alternative name
M–NH$_3$	azane	ammine
M–NH$_2$–NH$_2$	diazane-N diazane-κN	hydrazine-N
M(–NH$_2$)(–NH$_2$) (chelate)	diazane-N,N' diazane-$\kappa^2 N,N'$	hydrazine-N,N'
M–NH=NH	diazene-N	
M(–NH)(=NH) (chelate)	diazene-N,N' diazene-$\kappa^2 N,N'$	
M–NH$_2$–NH$_2$–M	μ-diazane-N,N' μ-diazane-$1\kappa N:2\kappa N'$	
M–NH=NH–M	μ-diazene-N,N'	
M–N$_3$H	hydrogen trinitride	hydrogen azide

The stereochemical information in the name of the complex is provided by the use of polyhedral symbols and the configuration indices (Sections I-10.5 and I-10.6.3 of Note 4a).

Note 4e. In most nitrogen hydride ligands the kappa convention is not significantly more useful than the 'donor atom symbol'.

HYDRIDES OF NITROGEN

Examples:

1. [Cu(NH$_3$)$_4$]SO$_4$
   tetraamminecopper(II)sulfate

2. [WBr$_2$(NH=NH)(Ph$_2$PCH$_2$CH$_2$PPh$_2$)$_2$]
   dibromo(diazene-*N*)bis[ethane-1,2-diylbis(diphenylphosphane-*P*)]tungsten(II)
   (see Note 4f)

3. [{Mn(η-C$_5$H$_5$)(CO)$_2$}$_2$(μ-NH=NH)]
   μ-diazene-1$\kappa N,N'$-bis[dicarbonyl(η-cyclopentadienyl)manganese(I)], or
   μ-diazene-1κN:2$\kappa N'$-bis[dicarbonyl(η-cyclopentadienyl)manganese(I)]

4.

   bis(μ-diazane-*N,N'*)bis[dichlorobis(dimethylphenylphosphane)ruthenium(II)], or
   bis(μ-diazane-1κN:2$\kappa N'$)bis[dichlorobis(dimethylphenylphosphane)ruthenium(II)]

5.

   (*OC*-6-12)-dichloro(diazane-*N,N'*)bis(triphenylphosphane)cobalt(1+)

II-4.5.3 Ligands formally derived from the nitrogen hydrides by the loss of one or more hydrons

These ligands have the same formulae as the corresponding anions. Their names are systematically derived from those of the corresponding anions by replacing the final e of the anion name by o. The point of attachment of the ligand is indicated by use of the kappa convention. When alternative derivations are possible, the choice is made according to Section II-4.5.1. Some simple bonding situations are given in Tables II-4.5, II-4.6, and II-4.7.

The ligand obtained by removing all the hydrogen atoms from diazane (hydrazine) may be called dinitrogen, dinitrido(2–), or dinitrido(4–) according to its formal charge. In the absence of other evidence the name dinitrogen is used (Section II-4.5.1).

Examples:

1. [MoCl(NH)(Ph$_2$PCH$_2$CH$_2$PPh$_2$)$_2$]
   azanediidochlorobis[ethane-1,2-diylbis(diphenylphosphane-*P*]molybdenum(III)

2. K$_2$[OsCl$_5$N]
   potassium pentachloronitridoosmate(2–)

3. [Mo(NNH$_2$)O(S$_2$CNMe$_2$)$_2$]
   (diazanediido-*N*)bis(*N,N'*-dimethyldithiocarbamato)oxomolybdenum(VI)

4. [Ir$_3$(μ_3-N)(SO$_4$)$_6$(H$_2$O)$_3$]$^{3-}$
   μ_3-nitridotris[aquabis(sulfato)iridate](3–)

Note 4f. The ditertiary phosphane ligand may also be named 1,2-bis(diphenylphosphino-*P*)ethane.

HYDRIDES OF NITROGEN

Table II-4.5 Names of the anionic ligands derived from ammonia (see Note 4g)

Bound ligand	Systematic name	Alternative name
M–NH$_2$	azanido	amido
M=NH	azanediido	imido
M≡N	nitrido	
M–NH$_2$–M	μ-azanido	μ-amido
M–NH–M	μ-azanediido	μ-imido
M–N=M	μ-nitrido	
M–NH(μ_3, bridging three M)	μ_3-azanediido	μ_3-imido
M–N (bridging three M)	μ_3-nitrido	
M–N (bridging four M)	μ_4-nitrido	

Table II-4.6 Names of anionic ligands derived from diazane (hydrazine)

Bound ligand	Systematic name	Alternative name
M–NH–NH$_2$	diazanido-N	hydrazido-N
M(NH–NH$_2$) chelate	diazanido-N,N'	hydrazido-N,N'
M=N–NH$_2$	diazanediido-N	hydrazinediido-N
M(NH–NH) chelate	diazanido(2–)-N, N' diazene-N,N' (Note 4h)	hydrazido(2–)-N
M(N–NH$_2$) chelate	diazanediido-N,N' diazanido(2–)-N,N'	hydrazinediido-N,N' hydrazido(2–)-N,N'
M–NH–NH$_2$–M	μ-diazanido-N,N'	
M–NH–NH–M	μ-diazene-N,N' (Note 4h)	
M=N–NH$_2$–M	μ-diazanediido-N,N'	
M–NH(NH$_2$)–M bridge	μ-diazanido-N	
M–N(NH$_2$)–M bridge	μ-diazanediido-N	

Note 4g. Only the simplest formal structures are shown. In transition metal compounds the M–N and N–N bond orders may be higher than shown.

HYDRIDES OF NITROGEN

Table II-4.7 Names for anionic ligands derived from diazene (see also Note 4h)

Bound ligand	Systematic name	Alternative name
M–N=N–H	diazenido-N	
M(–N–NH) (chelate)	diazenido-N,N'	
M–N(=NH)–M (bridging)	μ-diazenido-N	
M–N=NH–M	μ-diazenido-N,N'	

5. [CoH(N$_2$)(PPh$_3$)$_3$]
   (dinitrogen-N)hydridotris(triphenylphosphane)cobalt(I)

II-4.5.4 Ligands derived from the nitrogen hydride monocations by loss of one or more hydrons from the uncharged nitrogen centre

Ligands derived from the nitrogen hydride cations are named by adding the suffix -ido to the cation name (see Table II-4.8). The charge number is used to indicate the total formal charge on the ligand, including zero and one unit of negative charge.

Table II-4.8 Names of ligands derived from nitrogen hydride cations

Bound ligand	Systematic name	Alternative name
M–NH–NH$_3$$^+$	diazaniumido(0)	hydraziniumido(0)
M=N–NH$_3$$^+$	diazaniumido(1–)	hydraziniumido(1–)
[M–N(NH$_3$)(H)–M]$^+$	μ-diazaniumido(0)	μ-hydraziniumido(0)
[M–N(NH$_3$)–M]$^+$	μ-diazaniumido(1–)	μ-hydraziniumido(1–)
[M,M,M–N–NH$_3$]$^+$	μ_3-diazaniumido(1–)	μ_3-hydraziniumido(1–)
M–N=NH$_2$$^+$	diazeniumido(0) diazanido(2–) (Note 4i)	

Note 4h. Symmetrical diazanido(2–) and all diazanido(3–) ligands are named as derived from diazene (see Table II-4.4).

Note 4i. The ligand –N=NH$_2$ and other diazenium ligands are normally named as hydrazine derivatives (see Table II-4.6).

II-4.6 ORGANIC DERIVATIVES OF THE NITROGEN HYDRIDE LIGANDS

Organic derivatives are named as substitution products of the parents which have been named as described in Section II-4.1 to Section II-4.5. When the compound contains two contiguous nitrogen atoms they are designated by N^1 and N^2. In ligands where both nitrogen atoms carry no formal charge or when the nitrogen atoms carry equal formal charges, the number 1 is assigned to the nitrogen atom carrying most substituents. When each nitrogen atom has the same number of substituents, that with the substituent of lower alphabetical order is numbered 1. The metal atom does not count as a substituent for these purposes. In ligands which contain formally negatively charged nitrogen (see Section II-4.5), as well as neutral or positively charged nitrogen, the anionic nitrogen is designated by the number 1. If both nitrogen atoms carry formal negative charges, that carrying the greater negative charge has priority for number 1 (for more complex substitutions, see Chapters C-8 and C-9 of Note 4c).

Table II-4.9 Examples of linkages of substituted nitrogen hydride ligands

Bound ligand	Ligand name
M–N=N–C$_6$H$_5$	2-phenyldiazenido-N^1
M=N–N(CH$_3$)$_2$	2,2-dimethyldiazanido(2–)-N^1
M–N(CH$_3$)=N(C$_2$H$_5$)	1-ethyl-2-methyldiazene-N^2
M–N(CH$_3$)–N(C$_2$H$_5$)(C$_3$H$_7$)	1-ethyl-2-methyl-1-propyldiazanido-N^2
M–N(CH$_3$)–N(CH$_3$)$_2$H$^+$	1,2,2-trimethyldiazaniumido(0)-N^1
M–NH(CH$_3$)–N(CH$_3$)$_2$	1,1,2-trimethyldiazane-N^1
M–NH=N(CH$_3$)–M	μ-1-methyldiazene-N^1N^2

Examples:

1.
   ```
            P(C2H5)3
               |
      Cl——Pt——N≡N——Ph
               |
            P(C2H5)3
   ```
   trans-chloro(2-phenyldiazenido-N^1)bis--(triethylphosphane)platinum(II) (see Note 1b, p. 10)

2.
   ```
   [        P(C2H5)3          ]
            |    H
      Cl——Pt——N=N——Ph    Cl
            |
            P(C2H5)3
   ```
   trans-chloro(1-phenyldiazene-N^2)bis--(triethylphosphane)platinum(II)chloride

3. [WBr$_2${=N–N(CH$_3$)$_2$}(Ph$_2$PCH$_2$CH$_2$PPh$_2$)$_2$]
   dibromo[2,2-dimethyldiazanido(2–)-N^1]bis[ethane-1,2-diylbis(diphenylphosphane-P)]-tungsten(IV)

II-5 Inorganic Chain and Ring Compounds

CONTENTS

II-5.1 Introduction
II-5.2 Unbranched chain and monocyclic compounds
    II-5.2.1 Chain compounds
        II-5.2.1.1 General
        II-5.2.1.2 Choice of principal chain
        II-5.2.1.3 Numbering of chains
        II-5.2.1.4 Construction of the name
        II-5.2.1.5 Cations
        II-5.2.1.6 Anions
            II-5.2.1.6.1 Chain anions
            II-5.2.1.6.2 Anionic ligand name
    II-5.2.2 Monocyclic compounds
        II-5.2.2.1 Numbering of ring atoms
            II-5.2.2.1.1 Choice of position one
            II-5.2.2.1.2 Direction of numbering
            II-5.2.2.1.3 Procedure of numbering
        II-5.2.2.2 Cationic monocyclic ring compounds
        II-5.2.2.3 Anionic monocyclic ring compounds
II-5.3 Branched chain and polycyclic compounds
    II-5.3.1 Introduction
    II-5.3.2 The nodal descriptor
        II-5.3.2.1 Basic definitions
        II-5.3.2.2 The numbering of nodes
            II-5.3.2.2.1 General considerations
            II-5.3.2.2.2 Acyclic graphs
            II-5.3.2.2.3 Monocyclic graphs
            II-5.3.2.2.4 Polycyclic graphs
            II-5.3.2.2.5 Assemblies
        II-5.3.2.3 Descriptor
            II-5.3.2.3.1 Acyclic graphs
            II-5.3.2.3.2 Cyclic graphs
            II-5.3.2.3.3 Assemblies
    II-5.3.3 Construction of the name
        II-5.3.3.1 General considerations
        II-5.3.3.2 Ligands
        II-5.3.3.3 The name construction
            II-5.3.3.3.1 Branched acyclic compounds
            II-5.3.3.3.2 Polycyclic compounds

II-5.3.3.3.3 Assemblies
II-5.3.3.3.4 Cage compounds
II-5.3.3.3.5 Ionic species and ligands
II-5.3.3.3.6 Coordination compounds
II-5.4 Conclusion

II-5.1 INTRODUCTION

This chapter is concerned with a systematic approach for the naming of inorganic chain and ring compounds which has a long history (Note 5a). It is based on additive nomenclature and thus does not require any prior knowledge about the nature of bonds between the atoms. Although the method can be applied to all compounds, its use is intended for inorganic compounds which are mainly composed of atoms other than carbon.

A neutral chain compound is called a catena and a neutral cyclic compound a cycle. The corresponding cations are called catenium and cyclium and the anions catenate and cyclate. The number of atoms in the chain or in the ring is given by a descriptor [n] placed immediately before the term catena or cycle and preceded by a hyphen.

Specific rules are needed for the selection and numbering of the principal chain or ring in the molecule. The elements which constitute the chain or ring framework are listed in alphabetical order complete with their locants and are named modifying the element substituent group names given in Table VII of Note 5b by substituting the terminal -io by -y (see Table II-5.1). All atoms (including hydrogen) or groups of atoms which are not part of the chain or ring framework are named as ligands. They are also listed in alphabetical order before the cited sequence of the chain or ring atom terms.

II-5.2 UNBRANCHED CHAIN AND MONOCYCLIC COMPOUNDS

II-5.2.1 Chain compounds

II-5.2.1.1 General

In some cases it is convenient to treat the chain compound as unbranched. The molecules can then be described by defining a principal chain. Any branches attached to the principal chain are named as ligands. The number of atoms in the principal chain is given by a descriptor [n] placed immediately before the term catena and preceded by a hyphen. Thus, a six-membered chain is a -[6]catena. The chain length is defined as the longest chain of atoms in the molecule, disregarding terminal hydrogen atoms. Terminal hydrogen atoms are named as ligands (see Section II-5.2.1.4).

Example:
1. HS—S—S—SH
   1,4-dihydridotetrasulfy-[4]catena

Note 5a. The history of naming of inorganic chain and ring compounds has been reviewed by I. Haiduc in *The Chemistry of Inorganic Ring Systems* (R. Steudel, ed.), pp. 451–477, Elsevier, Amsterdam, 1992, and by W.H. Powell and T.E. Sloan, *Phosphorus, Sulfur, Silicon and Related Elements*, **41**, 183 (1989).

Note 5b. *Nomenclature of Inorganic Chemistry, Recommendations 1990*, Blackwell Scientific Publications, Oxford, 1990.

Table II-5.1 The names of the elements constituting the principal chain

Ac	actiny	Ge	germy*	Pr	praseodymy
Ag	argenty	Gd	gadoliny	Pt	platiny
Al	aluminy	H	hydrony*	Pu	plutony
Am	americy	He	hely	Ra	rady
Ar	argony	Hg	mercury	Rb	rubidy
As	arsy*	Ho	holmy	Re	rheny
At	astaty	I	iody	Rh	rhody
Au	aury	In	indy	Ru	rutheny
B	bory	Ir	iridy	S	sulfy*
Ba	bary	K	potassy	Sb	stiby
Be	berylly	Kr	kryptony	Sc	scandy
Bi	bismy*	La	lanthany	Se	seleny
Bk	berkely	Li	lithy	Si	sily*
Br	bromy	Lr	lawrency	Sm	samary
C	carby*	Lu	lutety	Sn	stanny
Ca	calcy	Md	mendelevy	Sr	stronty
Cd	cadmy	Mg	magnesy	T	trity
Ce	cery	Mn	mangany	Ta	tantaly
Cf	californy	Mo	molybdy	Tb	terby
Cl	chlory	N	azy	Tc	technety
Cm	cury	Na	sody	Te	tellury
Co	cobalty	Nb	nioby	Th	thory
Cr	chromy	Nd	neodymy	Ti	titany
Cs	caesy	Ne	neony	Tl	thally
Cu	cupry	Ni	nickely	Tm	thuly
D	deutery	No	nobely	U	urany
Dy	dysprosy	Np	neptuny	V	vanady
Er	erby	O	oxy	W	wolframy
Es	einsteiny	Os	osmy	Xe	xenony
Eu	europy	P	phosphy*	Y	yttry
F	fluory	Pa	protactiny	Yb	ytterby
Fe	ferry	Pb	plumby	Zn	zincy
Fm	fermy	Pd	pallady	Zr	zircony
Fr	francy	Pm	promethy		
Ga	gally	Po	polony		

* The names do not derive directly from the substituent group names.

II-5.2.1.2   *Choice of principal chain*

If a penultimate chain atom is bound to more than one atom different from hydrogen, the terminal atom chosen is the one first encountered in the element sequence in Table II-1.

Example:

1.
        NH$_2$
        |
HO—Si—
        |
        SH

Oxygen occurs first in the sequence and is preferred to sulfur and nitrogen. Therefore, O is a part of the chain.

If the terminal atoms of branched chains are identical, the choice of the principal chain is decided by comparison of the penultimate atoms or, if these also are identical, at the first point of difference moving inwards.

Example:

2.
$$H_3Si-\underline{O}-Si\begin{array}{c}S-SiH_3\\ |\\ |\\ Se-SiH_3\end{array}$$

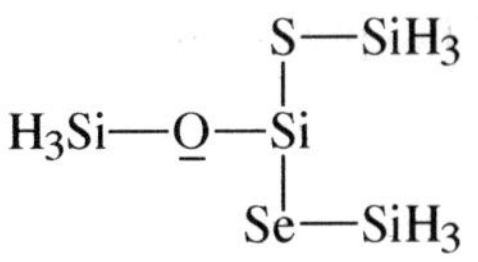

O is preferred to S and Se.

If the penultimate atom is bound to two identical atoms or the branches contain identical sequences of atoms but differ in the coordination numbers of some of the atoms, the principal chain should contain the element with higher coordination number.

Examples:

3.
$$H\underline{O}-\overset{\overset{O}{\|}}{\underset{H}{P}}-$$

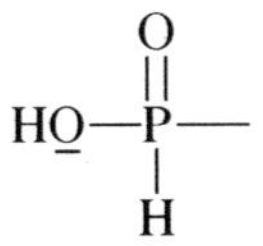

Oxygen with coordination number 2 is preferred to oxygen with coordination number 1.

4.
$$H_3Si-\overset{H}{\underset{H}{Si}}-\overset{}{\underset{\underset{Cl}{P}\diagdown Cl}{Si}}-\overset{\overset{O}{\|}}{\underset{Cl}{P}}-Cl$$

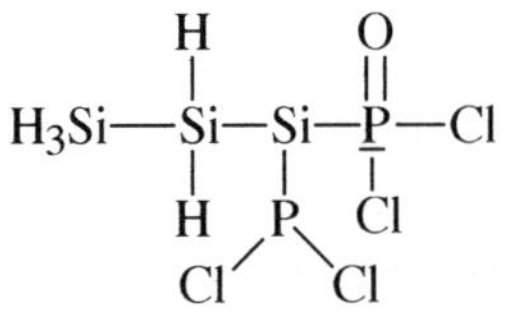

Phosphorus with coordination number 4 is preferred to phosphorus with coordination number 3.

II-5.2.1.3 Numbering of chains

The chain is numbered from that end which assigns a lower locant, at the first point of difference, to the elements first encountered in the sequence of Table II-1.

Examples:

1.
$$\overset{7}{C}-\overset{6}{Si}-\overset{5}{O}-\overset{4}{S}-\overset{3}{N}-\overset{2}{S}-\overset{1}{C}$$

The direction of numbering is governed by the choice of atom in position 2. S is preferred to Si.

2.
$$\overset{7}{N}-\overset{6}{N}-\overset{5}{Ge}-\overset{4}{Sn}-\overset{3}{Si}-\overset{2}{N}-\overset{1}{N}$$

The direction is governed by the choice of atom in position 3. Si is preferred to Ge.

In chains which are symmetrical with respect to the central atom(s), the chain is numbered from that end which assigns a lower locant, at the first point of difference, to the element with the highest coordination number. In the case of equal coordination numbers the ligands with donor atoms appearing first in Table II-1 have preference. In the case of polyatomic ligands the first differing atom appearing first in Table II-1 has preference.

Example:
3.
```
       7   6   5   4   3   2   1
                       O
                       ‖
       HO—NH—P—O—P—NH—OH
                |       |
                Cl      Cl
```

Points of difference are also atoms of an element in different oxidation states. The atom in the higher oxidation state is preferred to that in the lower oxidation state.

Example:
4.
```
       7    6    5   4   3    2    1
       C—Pb^II—C—O—C—Pb^IV—C
```

Pb^{IV} is preferred to Pb^{II}.

II-5.2.1.4   *Construction of the name*

The name of a principal chain is constructed by citation of a series of -y terms indicating the presence of each atom in the chain. The -y terms are arranged alphabetically and provided with locants and multiplying prefixes, di-, tri-, tetra-, *etc.*, as appropriate. All atoms or groups of atoms that are not considered as part of the principal chain (including hydrogen) are named as ligands according to coordination nomenclature principles and listed together with their locants in alphabetical order before the cited sequence of chain atom terms. Side chains consisting of common hydrocarbon substituents are denoted by customary substituent names, *e.g.* methyl.

Examples:
1.
```
       1   2   3   4   5   6   7   8
       HO—S—O—Se—O—S—S—OH
```

1,8-dihydrido-1,3,5,8-tetraoxy-4-seleny-2,6,7-trisulfy-[8]catena

2.
```
        6    5    4    3    2    1
        Cl—Si—Sn=Sn—Si—Cl
            H₂   H    H   H₂
```

2,2,3,4,5,5-hexahydrido-1,6-dichlory-2,5-disily-3,4-distanny-[6]catena

3.
```
         1    2    3    4   5    6    7    8
         H₃C—Si—Si—N—N—Si—Si—CH₃
              Me₂  H₂  H   H   H₂  Me₂
```

1,1,1,3,3,4,5,6,6,8,8,8-dodecahydrido-2,2,7,7-tetramethyl-4,5-diazy-1,8-dicarby-2,3,6-trisily-7-stanny-[8]catena

INORGANIC CHAIN AND RING COMPOUNDS

4.
$$\overset{1}{H_3C}-\overset{2}{\underset{H}{C}}=\overset{3}{N}-\overset{4}{\underset{H_2}{Si}}-\overset{5}{\underset{H}{N}}-\overset{6}{\underset{H}{C}}=\overset{7}{\underset{H}{C}}-\overset{8}{\underset{H}{N}}-\overset{9}{\underset{H_2}{Si}}-\overset{10}{\underset{H}{N}}-\overset{11}{C}\equiv\overset{12}{CH}$$

1,1,1,2,4,4,5,6,7,8,9,9,10,12-tetradecahydrido-3,5,8,10-tetraazy-1,2,6,7,11,12-hexacarby-4,9-disily-[12]catena

5.
$$\overset{1}{F}-\overset{2}{\underset{H_2}{Si}}-\overset{3}{\underset{H_2}{Si}}-\overset{4}{\underset{H}{N}}-\overset{5}{\underset{H}{N}}-\overset{6}{\underset{HBr}{Si}}-\overset{7}{SiH_3}$$

6-bromo-2,2,3,3,4,5,6,7,7,7-decahydrido-4,5-diazy-1-fluory-2,3,6,7-tetrasily-[7]catena

6.
$$\overset{1}{H_3Si}-\overset{2}{\underset{HF}{Si}}-\overset{3}{\underset{H}{N}}-\overset{4}{\underset{H}{N}}-\overset{5}{\underset{HBr}{Si}}-\overset{6}{SiH_3}$$

5-bromo-2-fluoro-1,1,1,2,3,4,5,6,6,6-decahydrido-3,4-diazy-1,2,5,6-tetrasily-[6]catena

7.
$$\overset{1}{Cl}-\overset{2}{\underset{H_2}{Si}}-\overset{3}{\underset{\underset{H}{|}}{\overset{\overset{CH_3}{|}}{Si}}}-\overset{4}{N}=\overset{5}{S}=\overset{6}{O}$$

2,2,3-trihydrido-3-methyl-4-azy-1-chlory-6-oxy-2,3-disily-5-sulfy-[6]catena

8.
$$\overset{1}{H_3Si}-\overset{2}{\underset{H_2}{Si}}-\overset{3}{O}-\overset{4}{\underset{|}{P}}-\overset{5}{O}-\overset{6}{\underset{|}{P}}-\overset{7}{O}-\overset{8}{\underset{H_2}{Si}}-\overset{9}{SiH_3}$$
$$\underset{1}{H_3Si}-\underset{2}{\underset{H_2}{Si}}-\underset{3}{S} \qquad SH$$

1,1,1,2,2,8,8,9,9,9-decahydrido-4-{2,2,3,3,3-pentahydrido-2,3-disily-1-sulfy-[3]caten-1--ato}-6-sulfanido-3,5,7-trioxy-4,6-diphosphy-1,2,8,9-tetrasily-[9]catena (see Note 1b, p. 10)

9.
$$\overset{1}{HO}-\overset{2}{\underset{\underset{O}{||}}{\overset{\overset{OH}{|}}{As}}}-\overset{3}{\underset{H_2}{C}}-\overset{4}{\underset{H_2}{C}}-\overset{5}{\underset{H}{N}}-\overset{6}{\underset{H}{N}}-\overset{7}{\underset{\underset{O}{||}}{C}}-\overset{8}{OH}$$

1,3,3,4,4,5,6,8-octahydrido-2-hydroxo-2,7-dioxo-2-arsy-5,6-diazy-3,4,7-tricarby-1,8--dioxy-[8]catena

INORGANIC CHAIN AND RING COMPOUNDS   II-5.2

10.
$$\overset{4\quad\;3\quad\;2\quad\;1}{\text{HS—S—Se—OH}}$$
with =O (up) and =O (down) on Se (position 2)

1,4-dihydrido-2,2-dioxo-1-oxy-2-seleny-3,4-disulfy-[4]catena

11.
$$\overset{1\quad\;2\quad\;3\quad\;4\quad\;5\quad\;6}{\text{HS—P—O—P—S—NH}_2}$$
with H below P at positions 2 and 4

1,2,4,6,6-pentahydrido-6-azy-3-oxy-2,4-diphosphy-1,5-disulfy-[6]catena

12.
$$\overset{1\quad\;2\quad\;3\quad\;4\quad\;5\quad\;6}{\text{HS—P—O—P—S—NH}_2}$$
with H_3 below P at position 2 and H below P at position 4

1,2,2,2,4,6,6-heptahydrido-6-azy-3-oxy-2,4-diphosphy-1,5-disulfy-[6]catena

13.
$$\overset{1\quad\;2\quad\;3\quad\quad 4\quad\;\;5\quad\;6\quad\;7\quad\;8\quad\;9}{\text{HO—As—C—C—N—N—C—N—NH}_2}$$
with OH above As, =O below As, H_2 below C (positions 3,4), H below N (positions 5,6), =O below C (position 7), H below N (position 8)

1,3,3,4,4,5,6,8,9,9-decahydrido-2-hydroxo-2,7-dioxo-2-arsy-5,6,8,9-tetraazy-3,4,7-
-tricarby-1-oxy-[9]catena

14.
$$\overset{1\quad\;2\quad\;3\quad\;4\quad\;5\quad\;6}{\text{HO—P—O—P—P—OH}}$$
with =O above P (positions 2,4,5), NH_2 below P (position 2), OH below P (position 4), SH below P (position 5)

2-amido-1,6-dihydrido-4-hydroxo-2,4,5-trioxo-5-sulfanido-1,3,6-trioxy-2,4,5-triphosphy-
-[6]catena

15.
$$\overset{1\quad\;2\quad\;3\quad\;4\quad\;5\quad\;6\quad\;7}{\text{HN=C—S—Sn—S—C}\equiv\text{N}}$$
with H below C (position 2) and H_2 below Sn (position 4)

1,2,4,4-tetrahydrido-1,7-diazy-2,6-dicarby-4-stanny(IV)-3,5-disulfy-[7]catena

When metal atoms are present in the chain, their oxidation states should be given, *i.e.* the oxidation number of the metal is indicated by a Roman numeral placed in parentheses immediately following the name of the element. For zero the symbol 0 is used. Examples are

INORGANIC CHAIN AND RING COMPOUNDS

chromy(II), chromy(III), chromy(0), *etc.* When used in conjunction with symbols the Roman numeral is placed as a right superscript (see Example 3 of Section II-5.2.1.6.1).

II-5.2.1.5 *Cations*

Chain cations are named catenium ions. The charge of the cation indicated by an Arabic numeral followed by the plus sign is placed in parenthesis following the ending -ium. The location of the charge in the backbone of the chain can be indicated by a locant before the ending -ium.

Examples:

1.
$$\overset{8}{HO}-\overset{7}{S}-\overset{6}{O}-\overset{5}{Se}-\overset{4}{O}-\overset{3}{S}-\overset{2}{\underset{F}{S^+}}-\overset{1}{F}$$

   2-fluoro-8-hydrido-1-fluory-4,6,8-trioxy-5-seleny-2,3,7-trisulfy-[8]catenium(1+) or
   2-fluoro-8-hydrido-1-fluory-4,6,8-trioxy-5-seleny-2,3,7-trisulfy-[8]caten-2-ium(1+)

2.
$$\left[\overset{1}{Me}-\overset{2}{\underset{Me_2}{N^+}}-\overset{3}{\underset{H_2}{Si}}-\overset{4}{S}-\overset{5}{\underset{H_2}{Si}}-\overset{6}{\underset{H}{N}}-\overset{7}{Me} \right] Br$$

   1,1,1,3,3,5,5,6,7,7,7-undecahydrido-2,2-dimethyl-2,6-diazy-1,7-dicarby-3,5-disily-4-sulfy--[7]catenium(1+)bromide or
   1,1,1,3,3,5,5,6,7,7,7-undecahydrido-2,2-dimethyl-2,6-diazy-1,7-dicarby-3,5-disily-4-sulfy--[7]caten-2-ium(1+) bromide

If the position of the positive charge of a catenium ion is uncertain, the compound is simply designated a catenium ion.

Examples:

3. $[H_2N-PPh_2-N-PPh_2-NH_2]Cl$
   1,1,5,5-tetrahydrido-2,2,4,4-tetraphenyl-1,3,5-triazy-2,4-diphosphy-[5]catenium(1+) chloride

4. $[Cl-P(NH_2)Cl-N-P(NH_2)Cl-Cl]Cl$
   2,4-diamido-2,4-dichloro-3-azy-1,5-dichlory-2,4-diphosphy-[5]catenium(1+) chloride

5.
$$\left[\overset{1}{Cl}-\overset{2}{\underset{NH_2}{\overset{Cl}{P}}}-\overset{3}{\underset{}{\overset{H}{N}}}-\overset{4}{\underset{}{\overset{H}{N}}}-\overset{5}{\underset{NH_2}{\overset{Br}{P}}}-\overset{6}{Cl} \right] Cl_2$$

   2,5-diamido-5-bromo-2-chloro-3,4-dihydrido-3,4-diazy-1,6-dichlory-2,5-diphosphy--[6]catenium(2+) dichloride

II-5.2.1.6 Anions

II-5.2.1.6.1 *Chain anions.* Chain anions are named catenate ions. The charge of the anion, indicated by an Arabic numeral followed by the minus sign, is placed in parentheses following the ending -ate. The location of the charge in the backbone of the chain may be indicated by a locant before the ending -ate.

Examples:

1.

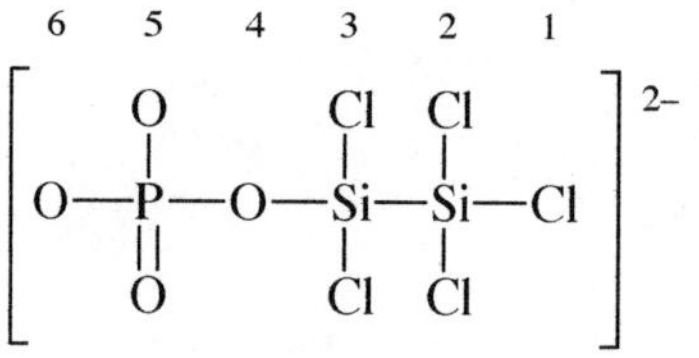

$$\begin{bmatrix} & & & & & \text{Cl} & \text{Cl} & \\ & \text{O} & & & & | & | & \\ \text{O}-\text{P}-\text{O}-\text{Si}-\text{Si}-\text{Cl} \\ & \| & & & & | & | & \\ & \text{O} & & & & \text{Cl} & \text{Cl} & \end{bmatrix}^{2-}$$

positions: 6 5 4 3 2 1

2,2,3,3-tetrachloro-5,5-dioxo-1-chlory-4,6-dioxy-5-phosphy-2,3-disily-[6]catenate(2−)

2.

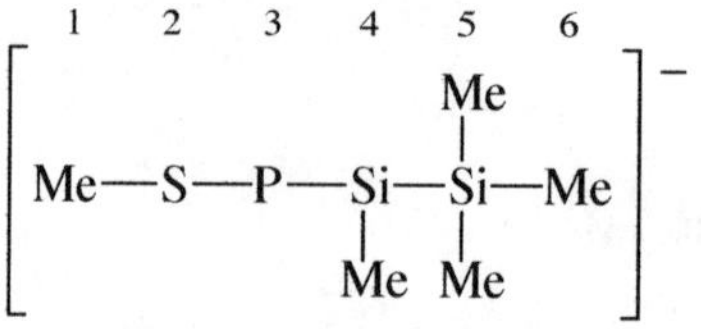

$$\begin{bmatrix} & & & & \text{Me} & \\ & & & & | & \\ \text{Me}-\text{S}-\text{P}-\text{Si}-\text{Si}-\text{Me} \\ & & & | & | & \\ & & & \text{Me} & \text{Me} \end{bmatrix}^{-}$$

positions: 1 2 3 4 5 6

1,1,1,6,6,6-hexahydrido-4,5,5-trimethyl-4-azy-1,6-dicarby-3-phosphy-5-sily-2-sulfy--[6]caten-3-ate(1−)

3.

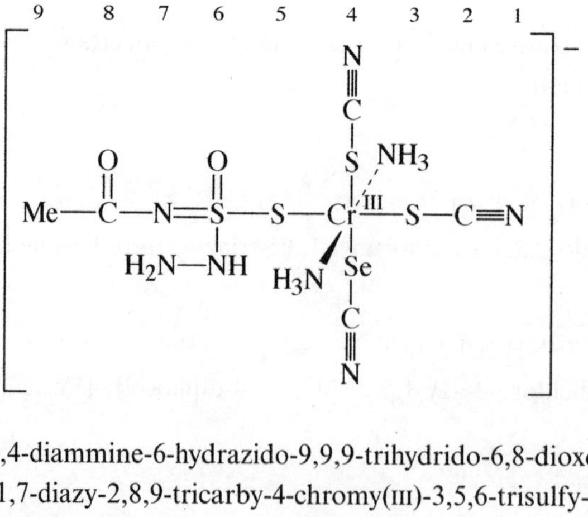

$$\begin{bmatrix} \text{structure with Cr(III) center} \end{bmatrix}^{-}$$

positions: 9 8 7 6 5 4 3 2 1

4,4-diammine-6-hydrazido-9,9,9-trihydrido-6,8-dioxo-4-selenocyanato-4-thiocyanato--1,7-diazy-2,8,9-tricarby-4-chromy(III)-3,5,6-trisulfy-[9]catenate(1−), or
4,4-diammine-4-(3-azy-2-carby-1-seleny-[3]caten-1-ato)-4-(3-azy-2-carby-1-sulfy--[3]caten-1-ato)-6-hydrazido-9,9,9-trihydrido-6,8-dioxo-1,7-diazy-2,8,9-tricarby-4--chromy(III)-3,5,6-trisulfy-[9]caten-4-ate(1−)

If the position of the negative charge of a catenate ion is uncertain, it is simply designated a catenate ion.

INORGANIC CHAIN AND RING COMPOUNDS II-5.2

Example:
4.

$$Na^+ \left[\begin{array}{c} \overset{1}{}\overset{2}{}\overset{3}{}\overset{4}{}\overset{5}{}\overset{6}{}\overset{7}{} \\ Me-\underset{\underset{O}{\|}}{\overset{\overset{O}{|}}{P}}-O-\underset{Me}{\overset{}{P}}-O-\underset{Me}{\overset{}{P}}-Me \end{array} \right]^-$$

sodium 1,1,1,7,7,7-hexahydrido-4,6-dimethyl-2,2-dioxo-1,7-dicarby-3,5-dioxy-2,4,6--triphosphy-[7]catenate

II-5.2.1.6.2 *Anionic ligand name.* Ligands derived from catenas by the loss of hydrons are given the ending -ato. The location of the charge may be indicated with a locant placed before the ending.

Examples:
1. $[H_3Si-SiH_2-S]^-$
   2,2,3,3,3-pentahydrido-2,3-disily-1-sulfy-[3]caten-1-ato

2. $$\left[HS-S-\overset{..}{\underset{..}{N}}-SeH \right]^-$$

   1,4-dihydrido-3-azy-4-seleny-1,2-disulfy-[4]caten-3-ato

II-5.2.2 **Monocyclic compounds**

The following nomenclature provides names for inorganic ring compounds consisting mainly of elements other than carbon. Ring systems containing two or more consecutive carbon atoms are usually named according to the rules of Note 5c.

A neutral monocyclic ring compound is called a cycle, an anion a cyclate ion, and a cation a cyclium ion. The number of atoms in the ring is given by an indicator [*n*] placed immediately before the term cycle, cyclate, or cyclium, and preceded by a hyphen. Thus, a six-membered ring is a -[6]cycle. In order to emphasize the cyclic structure with the descriptor it is acceptable to use a zero before the Arabic number, *i.e.* -[06]cycle (see Section II-5.3.2.3.2).

The names of derivatives of cycles are formed from the cycle name using coordination nomenclature, see Chapters I-7 and I-10 of Note 5b. All ligands, including hydrogen, are named by coordination nomenclature and listed in alphabetical order.

II-5.2.2.1 *Numbering of ring atoms*

II-5.2.2.1.1 *Choice of position one.*

(a) The elements which constitute the ring are named according to Table VII of Note 5b by substituting the terminal -io by -y (see Table II-5.1) and listed in alphabetical order.
(b) Numbering starts at the atom which is listed first in Table II-1.

Note 5c. *Nomenclature of Organic Chemistry*, Pergamon Press, Oxford, 1979 (see also *A Guide to IUPAC Nomenclature of Organic Compounds, Recommendations 1993*, Blackwell Scientific Publications, Oxford, 1994).

(c) If two of more atoms of highest seniority are ring members, the one with the neighbour which follows next in Table II-1 is the starting atom. If this second operation does not define a unique starting atom, neighbours once removed are sequentially considered, until a unique starting atom has been found or shown not to exist.

(d) If the above criteria fail to define a unique starting atom, it is determined by consideration of the coordination numbers of those ring atoms which have the highest seniority in Table II-1. The starting atom is the one with the highest coordination number. If needed, the coordination numbers of the neighbouring atoms are sequentially considered, until a unique starting atom has been found or shown not to exist.

(e) If the application of the above criteria identifies two or more equal contenders as starting atoms, a choice must be made between them. Priority is given to the atom whose ligand comes first alphabetically.

(f) If all ligands are equal, the ligands on neighbours once removed are considered sequentially, until a unique starting atom has been found. If no unique starting atom is found by this method, the symmetry of the ring and its ligands is such that an arbitrary choice will lead to a unique name. The starting atom may be arbitrarily chosen from the equal contenders for priority.

II-5.2.2.1.2 *Direction of numbering.* This is determined by application of the following criteria in order until a decision is reached:

(a) The direction of numbering leads from position 1 to whichever neighbouring atom is listed first in Table II-1.
(b) If both neighbouring atoms are the same, the pair of next nearest neighbours is compared, until a unique atom first in Table II-1 is identified.

Examples:
1. Unique numbering.

2. Alternative numbering, as shown in parentheses.

(c) When the direction of numbering cannot be determined by considering the ring atoms it is determined by considering the ligands. The direction of numbering leads from the starting atom to a neighbour with a ligand.
(d) If both have ligands it leads to that whose ligand comes first alphabetically.

(e) If the ligands are equal the ligands on next nearest neighbours are considered and so on until a unique direction is attained. If none is found the symmetry of ring and ligands leads to a unique name independently of the direction.

Examples:

1.

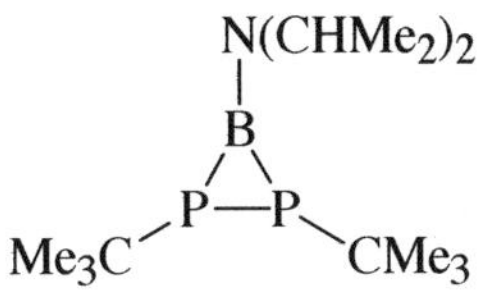

$$\underset{Me_3C}{}\overset{\overset{\displaystyle N(CHMe_2)_2}{|}}{\underset{}{B}}\overset{}{\underset{P\!-\!P}{\diagup\quad\diagdown}}\underset{CMe_3}{}$$

1,2-di-*tert*-butyl-3-diisopropylamino-3-bory-1,2-diphosphy-[3]cycle

2.

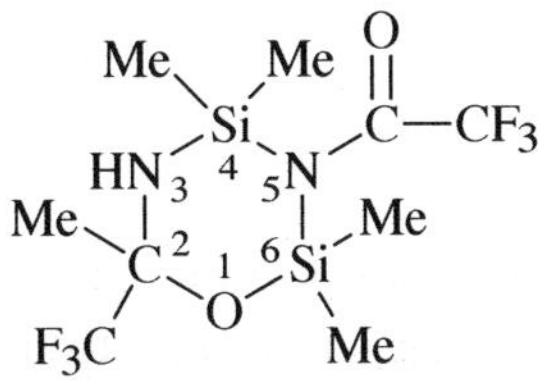

$$\begin{array}{c} Me \quad Me \quad O \\ \diagdown Si \diagup \quad \| \\ HN_{\,3}\;{}^4{}_5N\;{-}C{-}CF_3 \\ Me\diagdown|_{\,2}\quad\;{}_6|\diagup Me \\ C{-}{}_1{-}Si \\ \diagup\quad O\quad \diagdown \\ F_3C\qquad\quad Me \end{array}$$

3-hydrido-2,4,4,6,6-pentamethyl-5-trifluoroacetyl-2-trifluoromethyl-3,5-diazy-2-carby-1-oxy-4,6-disily-[6]cycle

II-5.2.2.1.3 *Procedure of numbering.* In order to facilitate the numbering, the ring members in a given system can be alphabetically labelled which indicates their relative position in the sequence of Table II-1.

In the ring

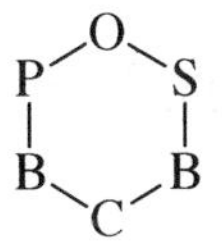

$$\begin{array}{c} O \\ P\diagup\;\diagdown S \\ | \qquad | \\ B\diagdown\;\diagup B \\ C \end{array}$$

the relative positions in the sequence of elements are as follows (see Section II-5.2.2.1.2):

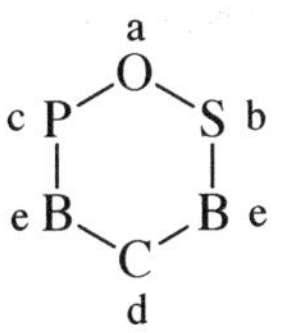

$$\begin{array}{c} a \\ O \\ c\,P\diagup\;\diagdown S\,b \\ |\qquad\;| \\ e\,B\diagdown\;\diagup B\,e \\ C \\ d \end{array}$$

For the numbering, that direction is chosen which gives the initial sequence of the letters earliest in the alphabet. The resulting sequence a,b,e,d,e,c leads to the numbering

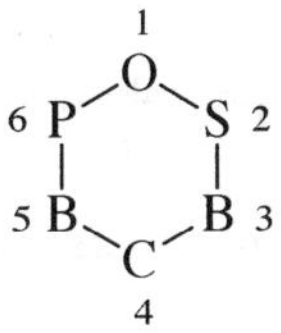

$$\begin{array}{c} 1 \\ O \\ 6\,P\diagup\;\diagdown S\,2 \\ |\qquad\;| \\ 5\,B\diagdown\;\diagup B\,3 \\ C \\ 4 \end{array}$$

If the initial atom in the ring has neighbours of the same kind on both sides, that sequence is preferred which contains the second or subsequent sequence of the letters earliest in the alphabet.

Thus in

the sequence is a,b,c,d,e,b and not a,b,e,d,c,b.

II-5.2.2.2   *Cationic monocyclic ring compounds*

Cyclic inorganic cations are named in the same way as the neutral ring systems, but instead of cycle the suffix cyclium is used. A locant for the charge may be inserted before -ium.

Example:
1.

1,1-dimethyl-3,3,5,5-tetraphenyl-2,4,6-triazy-3,5-diphosphy-1-sulfy-[6]cyclium bromide

II-5.2.2.3   *Anionic monocyclic ring compounds*

Cyclic inorganic anions are named in the same way as the neutral ring systems, but instead of cycle the suffix cyclate is used. A locant for charge may be inserted before -ate.

Examples:
1.

ammonium 3,3,5,5-tetraamido-1,1-dioxo-2,4,6-triazy-3,5-diphosphy-1-sulfy-[6]cyclate

When an anionic monocyclic ring acts as a ligand, its name is modified by replacing the ending -ate by -ato.

2.

tetrakis(3,3,5,5-tetraamido-1,1-dioxo-2,4,6-triazy-3,5-diphosphy-1-sulfy-[6]cyclato)uranium(IV)

II-5.3 BRANCHED CHAIN AND POLYCYCLIC COMPOUNDS

II-5.3.1 Introduction

The method described above can conveniently be extended to chain compounds containing more complicated branched structures, to polycyclic compounds, and to mixed chain and ring compounds by use of the general nodal descriptor (Note 5d).

Nodal nomenclature was originally developed to solve problems encountered in naming complicated organic molecules. It is based on the description of the structure of a molecule in terms of its graph. Each graph is thought to be composed of one or more modules which are defined as separate entities during the numbering and naming of the graph assembly. Each module in turn is composed of nodes that are the simplest units in the graph and represent either a single atom or a group of atoms (contraction nodes).

The general nodal nomenclature defines rules by which the nodes of the graph are arranged and numbered. It differs from the standard organic nomenclature because a unique numerical locant has to be provided for every node in the graph. At the most general level of treatment the nodal nomenclature does not specify the nature of the nodes or the bonding between them. Thus, it is not necessary to assume any specific coordination number for any node. Therefore nodal principles can equally well be applied for both the substitutive nomenclature used in organic chemistry (Note 5e) and for the coordination nomenclature common in inorganic chemistry (Note 5a). Nodal nomenclature is also shown to be useful when computerizing chemical nomenclature (Note 5f).

In the following, the basic rationale in the constructing of the nodal descriptor is summarized according to the concepts developed by Lozac'h, Goodson, and Powell (Note 5d) followed by their special application to branched chain and polycyclic inorganic systems.

Note 5d. N. Lozac'h, A.L. Goodson, and W.H. Powell, *Angew. Chem.*, **91**, 951 (1979); *Angew. Chem., Int. Ed. Engl.*, **18**, 887 (1979).
Note 5e. N. Lozac'h and A.L. Goodson, *Angew. Chem.*, **96**, 13 (1984); *Angew. Chem., Int. Ed. Engl.*, **23**, 33 (1984); D.J. Polton, *J.Chem.Inf.Comput.Sci.*, **32**, 430 (1992).
Note 5f. D.J. Polton, *Chemical Nomenclature and the Computer*, Research Studies Press, 1993.

II-5.3.2 The nodal descriptor

II-5.3.2.1 Basic definitions

The modules of the graph (or the graph itself if there is only one module in the structure) can be acyclic, monocyclic or polycyclic.

(a) An *acyclic graph* (module) is an unbranched chain of nodes, or two or more unbranched chains of nodes connected to each other without formation of a cyclic structure.
(b) A *monocyclic graph* (module) is constructed when one end of the chain of nodes is attached to the other end forming only one ring into the structure.
(c) A *polycyclic graph* (module) consists of a monocyclic ring of nodes and one or more bridges which are valence bonds or chains of nodes connecting the nodes of the ring and/or other bridges in the polycyclic system.
(d) If a structure graph is composed of more than one cyclic module or of at least one acyclic and one cyclic module, it is called an *assembly*.

II-5.3.2.2 The numbering of nodes

II-5.3.2.2.1 *General considerations*. Each node in the graph needs to be numbered. The general nodal nomenclature (Note 5d) provides detailed instructions on the relative seniority of the nodes regardless of their chemical identity. The chemical nature of the nodes is taken into consideration only to distinguish between two otherwise equivalent numberings.

This is also the rationale to be followed in naming more complicated inorganic ring and chain compounds. The seniority is decided as far as possible using basic rules of general nodal nomenclature. The priority rules derived for the simple chain and monocyclic compounds (see Section II-5.2) which take into account the chemical nature of the elements according to the element seniority sequence of Table II-1 are considered only when ambiguity arises.

II-5.3.2.2.2 *Acyclic graphs*. The acyclic graph is characterized by a main chain which is defined as the longest unbranched chain of nodes. Terminal hydrogen atoms are not considered as a part of the chain. They are always named as ligands. If the penultimate atom is bound to more than one atom different from hydrogen, the terminal atom is chosen as the one first encountered in the element seniority sequence (see Table II-1). All other chains in the acyclic graph are defined as branches.

If there is more than one unbranched chain of the greatest length, the main chain is the one having the longest branch attached to it. The nodes in the main chain are numbered from one end to another so that the lowest possible locant is given to this branch of highest priority (Note 5g). The branches are numbered in the order of their seniority (decreasing length) beginning from the node attached to the part of the graph already numbered.

Note 5g. Note that this practice is different in the chain nomenclature (Section II-5.2.2) where only the main chain is included in the basic framework and the branches are named as ligands. It is then the element seniority sequence (Table II-1) which governs the choice of the main chain.

Examples:

1.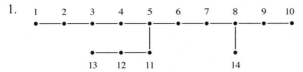

2.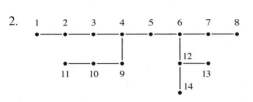

In this case there are two possible main chains (1–2–3–4–5–6–7–8 and 11–10–9–4–5–6–7–8). The choice has to be made considering the element seniority sequence.

3.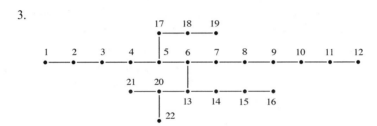

If alternative numberings remain, the choice is made by comparing the locants of the branches term by term, in the order that the branches are numbered, and selecting the lowest possible locants. Should any ambiguity still remain, it is resolved by considering the element seniority sequence.

Examples:

4.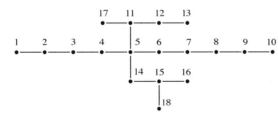

5.
$$\begin{array}{c}
{}_{11}N \\
\vertiii{} \\
{}_{10}C \\
{}_{}\!\!\!\!\!\overset{123}{H_2N-S-S}\!\!\!-\overset{4}{C}\!\!-\overset{567}{S-S-NH_2} \\
{}_8 S \\
{}_9 NH_2
\end{array}$$

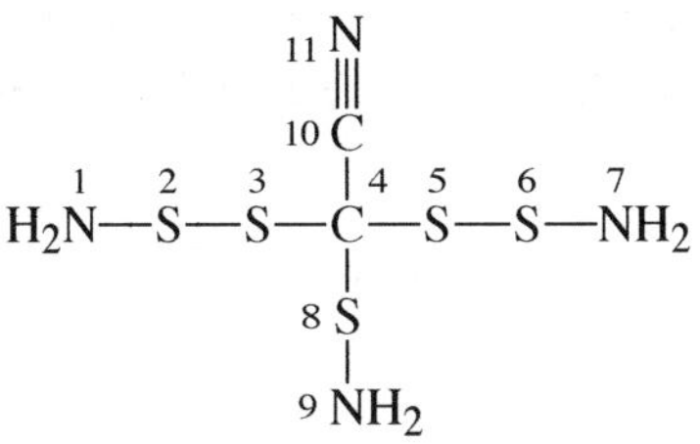

The principal chain is seven atoms long and its atoms are numbered starting from either of the nitrogen atoms of the NH$_2$ groups. Since terminal hydrogen atoms are not considered as a part of the nodal framework, the two branches connected with the main chain are of equal length. However, S is senior to C in the element seniority sequence and thus the branch containing S is numbered first.

II-5.3.2.2.3 *Monocyclic graphs.* The starting point and the direction of the numbering of the atoms in the monocyclic compound is determined by considering the chemical nature of the atoms constituting the ring or of the ligands attached to the ring according to the element seniority sequence (Table II-1).

II-5.3.2.2.4 *Polycyclic graphs.* A polycyclic graph is characterized by a main ring. Chains attached from both ends to the main ring are called bridges. The main ring in a polycyclic graph is defined as a monocyclic ring with largest number of nodes. The main bridge is defined as a longest chain of nodes both ends of which are attached to the main ring. All other bridges are called secondary bridges. Note that a bridge can also be a bond in which case it contains no nodes.

The numbering of the main ring in the polycyclic graph begins at one of the bridgeheads (the node where the main bridge is attached to the main ring) and proceeds in the direction to give a lowest possible locant for the other bridgehead. The bridges are numbered sequentially after the main ring in the order of their seniority, beginning always with the longest bridge connected to nodes of the graph previously numbered. Should there be two or more bridges of equal length, the one with lower locants has the highest seniority. The numbering of each bridge begins with the node connected to a node with a lower locant.

Examples:
1.
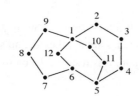

The main ring has nine nodes. The main bridge is the chain of two nodes connecting nodes 1 and 5.

2.
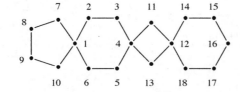

Should there be two or more monocyclic rings having the largest number of nodes, the main ring is selected as the one having a main bridge with the greater number of nodes.

3.

The atoms 1 and 4 are connected by a bond and thus the main bridge contains no nodes.

INORGANIC CHAIN AND RING COMPOUNDS

II-5.3.2.2.5 *Assemblies.* The definite numbering of the assemblies is carried out by numbering initially each module as if it were an isolated graph. The principal module and the seniority of the modules in assemblies are determined according to the successive application of the following criteria: (a) largest number of nodes; (b) the cyclic module preferred to an acyclic one; (c) largest number of rings or branches; (d) the descriptor with the preferred Arabic numeral at the first difference: if the first difference corresponds to a chain or a bridge length, the preferred numeral is higher; if the first difference appears in a locant (superscript), the preferred numeral is lower.

The original numbering is retained for the principal module. The numbering of nodes in all other modules is modified sequentially, beginning with a module adjoining the principal module (Note 5d). The renumbering is carried out by adding to each original locant a number equal to the total number of nodes in other modules that have already been assigned definitive locants.

Example:

1.

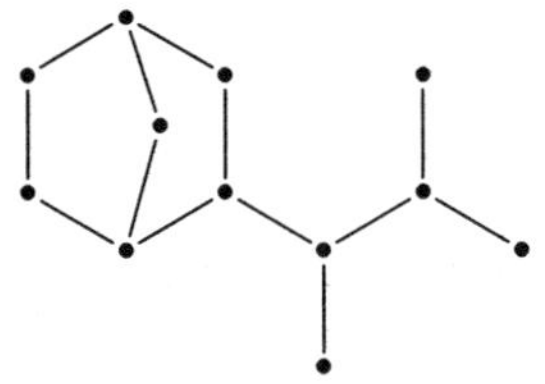

The molecule consists of one cyclic and one acyclic module which are initially numbered separately as follows:

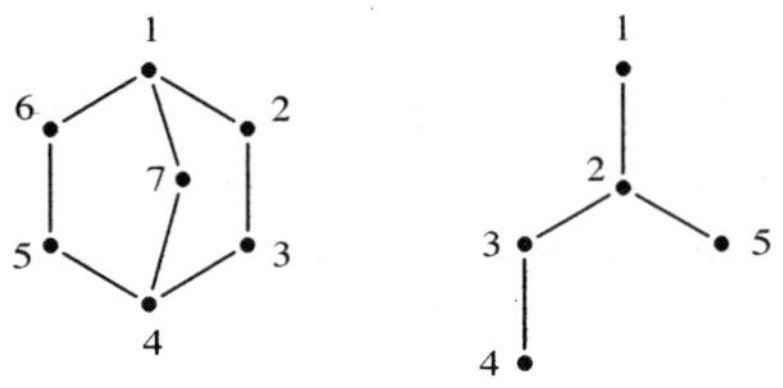

The cyclic module has preference over the acyclic one and is thus the principal module in the graph. The numbering in the acyclic module is modified by adding 7 to each locant. The final definitive numbering is accordingly the following:

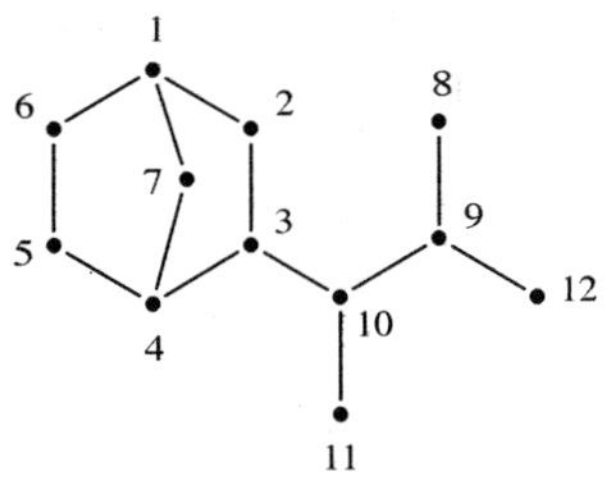

In some cases there are several renumbering schemes possible for the given assembly. The correct choice is made on the basis of a module seniority graph (Note 5d). In the case of any remaining ambiguity the correct numbering has the lowest locants defining the attachments of the modules. If needed, the final resolution is made by considering the element seniority sequence.

Example:
2.

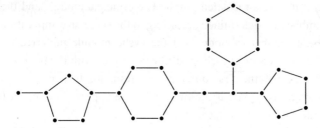

The graph consists of the following modules:

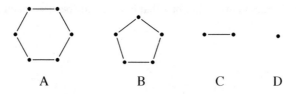

They are assigned module seniority descriptors A–D in the order of the decreasing seniority. In effect, the letters A–D represent contraction nodes. Thus the module seniority graph of the molecule can be written as follows:

$$\begin{array}{c} A \\ | \\ D-B-A-C-B \end{array}$$

The modules are now numbered using the module seniority numbers. They begin with the principal module and proceed through the chain of modules in order of their decreasing seniority. The rules for assigning module seniority numbers are similar, but not analogous, to the numbering of nodes in acyclic graphs (Note 5d):

$$\begin{array}{c} 1 \\ A \\ \overset{543}{D-B-A} \overset{|26}{-C-B} \end{array}$$

The definitive numbering of the nodes in the whole assembly is now carried out in the order of the module seniority numbers (for more details in the definitive numbering of the assemblies, see Note 5d).

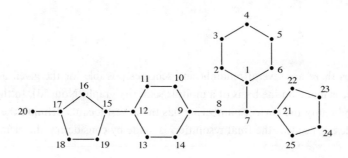

II-5.3.2.3 *Descriptor*

The unambiguous description of the graph is provided by a descriptor. It is a numerical portion in the name and is placed in square brackets.

II-5.3.2.3.1 *Acyclic graphs.* The descriptor of the acyclic graphs is constructed as follows:

(a) The descriptor of the acyclic graphs begins with an Arabic numeral indicating the number of the nodes in the main chain (Note 5h). This portion of the descriptor ends with a full stop.
(b) The full stop is followed by positive Arabic numerals indicating the number of nodes in each branch cited in the order of their seniority.
(c) A superscript locant for each branch denotes the node in the part of the graph already numbered to which the branch is attached.

When it is possible to write alternative descriptors for the graph the correct one has the lowest superscript number at the first point of difference on term by term comparison.

Examples:

1.

descriptor: [7]

2.
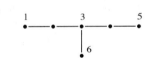

descriptor: [5.1^3]

3.
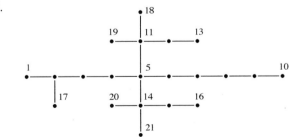

descriptor: [10.3^{5}3^{5}1^{2}1^{11}1^{11}1^{14}1^{14}]

Note 5h. If the nodal framework consists only of the main chain with all branches named as ligands, the nodal descriptor is identical with the indicator [*n*] used for the simple inorganic chain compounds.

II-5.3.2.3.2 *Cyclic graphs.* The descriptor of the monocyclic graphs is an arabic numeral indicating the number of the nodes in the ring. The numeral is preceded by zero which indicates the presence of the ring (Note 5i).

Example:
1.

descriptor: [06]

The descriptor for polycyclic graphs is constructed in a manner analogous to that for branched graphs.

(a) The descriptor begins with a zero indicating the presence of a ring followed by an arabic numeral indicating the number of nodes in the main ring. This portion of the descriptor ends with a full stop.
(b) The full stop is followed by Arabic numerals indicating the number of nodes in each bridge cited in the order of their seniority.
(c) A pair of superscript locants for each bridge numeral, separated by a comma and cited in increasing numerical order, denotes the nodes in the part of the graph already numbered to which each bridge is attached.

Should there be a choice for the main ring, main bridge, starting point and/or direction of numbering in the polycyclic graph, the arabic numerals denoting the lengths and positions of the bridges in the various alternative descriptors are compared term by term in the order they appear. The descriptor with the preferred Arabic numeral at the first difference is chosen (in case of the bridge length, the preferred numeral is higher; in case of the locant, lower).

Examples:
2.

descriptor: [07.1^{1,4}]

Note 5i. In the nomenclature for monocyclic compounds (See Section II-5.2.3) the zero is not a required part of the descriptor; its use is, however, permissible if the cyclic nature of the compound needs to be emphasized.

INORGANIC CHAIN AND RING COMPOUNDS

3.

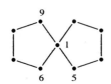

descriptor: [05.4^{1,1}]

4.

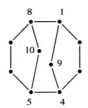

descriptor: [08.1^{1,4}1^{5,8}]

5.

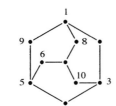

descriptor: [08.1^{1,5}1^{3,7}]

6.

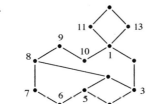

descriptor: [010.3^{1,1}1^{3,5}0^{4,8}]

II-5.3.2.3.3 *Assemblies.* The assembly descriptor consists of square brackets enclosing, in the order of their seniority, the nodal descriptors of each module in parentheses and cited as if they were isolated graphs. Between the descriptors of the modules the locants of the nodes linking the modules are indicated. These locants are separated by a colon and are referred to by their definitive sequential numbering obtained when considering the whole assembly.

Examples:
1.

descriptor: [(06)1:7(4)10:11(05)]

INORGANIC CHAIN AND RING COMPOUNDS

2.

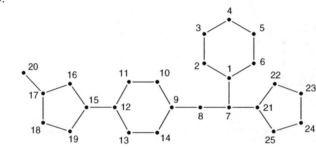

descriptor: [(06.1^{1,4})1:9(06.1^{8,11})]

The descriptors for both modules are obtained as if the module were an isolated graph. The highest node in the principal module has the locant seven. The lowest locant (eight) of the second module is at the bridgehead as required by the numbering rules and not at the node where the modules are attached together. Thus it is possible to infer from the descriptor that the two parts of the molecule are identical, but attached to each other at different respective locations.

3.

descriptor:[(06)1:7(2):8:9(06)12:15(05)17:20(1)7:21(05)]

II-5.3.3 Construction of the name

II-5.3.3.1 General considerations

When naming the compound, it is important to decide which parts of the molecule should be included in the nodal skeleton. All other parts will be named as ligands.

If the nodal skeleton consists only of the main chain or of the main ring, the compounds are named following the nomenclature derived for simple chain and monocyclic compounds (see Section II-5.2).

If the nodal skeleton is composed of a branched acyclic framework, the compound is called catena preceded by a multiplicative prefix di-, tri-, *etc.* to indicate the number of branches in the molecule. The cationic species are called catenium and the anionic species catenate.

In an analogous manner a neutral polycyclic compound is called cycle preceded by a multiplicative prefix di-, tri-, *etc.* to indicate the number of rings in the molecule. The cationic species are called cyclium and the anionic species cyclate.

If there are both cyclic and acyclic modules in the molecule the compound is named as catenacycle (the names are cited in alphabetic order) with both parts of the name preceded by appropriate multiplicative prefixes.

Examples:

1.

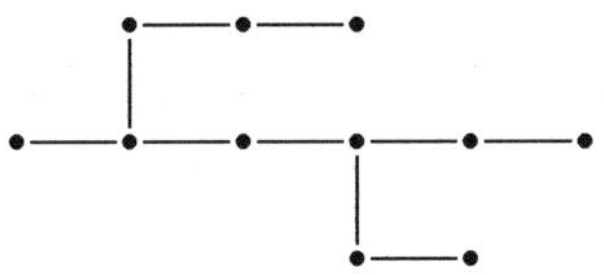

   tricatena

2.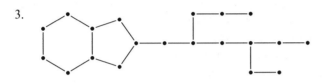

   dicycle

3.

   tricatenadicycle

Since the nodal descriptor only indicates the geometrical arrangement of the atoms, it is necessary to state unambiguously the chemical identity of each node. Following the practice introduced for the inorganic chain and monocyclic compounds the skeletal atoms forming the nodal framework are named modifying the substituent group names given in Table II-5.1. All atoms or groups of atoms which are not considered as part of the polycyclic framework are named as ligands.

II-5.3.3.2   *Ligands*

In principle all atoms in the molecule can be treated as a part of the nodal framework. The resulting names will, however, become too cumbersome to be practical. Therefore it is preferable to name some atoms or groups of atoms as ligands to the nodal skeleton.

There are atoms or groups of atoms which are usually considered as ligands in coordination compounds. Some examples of typical ligands are given in Table II-5.2.

Table II-5.2 Examples of typical ligands (Note 5j)

Ligand	Name	Ligand	Name
F^-	fluoro	CH_3	methyl
Cl^-	chloro	CH_3CH_2	ethyl
Br^-	bromo	$CH_3CH_2CH_2$	propyl
I^-	iodo	$(CH_3)_3C$	1,1-dimethylethyl
CO	carbonyl	C_5H_5	cyclopentadienyl
NCO^-	cyanato	C_6H_5	phenyl
Ph_3P	triphenylphosphane		

Note 5j. See Tables I-10.1, I-10.2, I-10.3, I-10.4 and Section I-10.9 of Note 5b for additional examples.

For a more detailed treatment, see Chapter I-10 of Note 5b. It is also permissible to name any part of the molecule as a ligand if it is considered convenient in specific situations.

There are, however, two restrictions in the use of ligands in nodal frameworks:

(a) All ligands must be monodentate.
(b) The main chain in acyclic modules can only be terminated by a ligand if the ligand is either a hydrogen atom or an organic fragment.

II-5.3.3.3 *The name construction*

The name consists of the following parts:

(a) Ligands listed in alphabetic order complete with their definitive sequential locants.
(b) Skeletal atoms forming the nodal framework listed in alphabetic order together with their definitive sequential locants when needed.
(c) The nodal descriptor in square brackets.
(d) The designator catena, cycle, or catenacycle for neutral molecules, catenium, cyclium, or catenacyclium for cations, and catenate, cyclate, or catenacyclate for anions, preceded by appropriate multiplicative prefixes.

II-5.3.3.3.1 *Branched acyclic compounds*

Examples:
1.

$$\begin{array}{ccccccc} 7 & 6 & 5 & 4 & 3 & 2 & 1 \\ H & H & H & H & H & H & \\ | & | & | & | & | & | & \\ H-Si-Si-Si-Si-Si-P&=&O \\ | & | & | & | & | & | & \\ H & H & H & {}_{}P^8 & H & H & \\ & & & {}_{10}Cl\phantom{_{0}} & Cl\,9 & & \end{array}$$

2,2,3,3,4,5,5,6,6,7,7,7-dodecahydrido-9,10-dichlory-1-oxy-2,8-diphosphy-3,4,5,6,7--pentasily-[7.2^4]dicatena

It might be preferable to emphasize in the name that the two chlorine atoms bound to P(8) are identical. The name would thus be: 8,8-dichloro-2,2,3,3,4,5,5,6,6,7,7,7-dodecahydrido-1-oxy--2,8-diphosphy-3,4,5,6,7-pentasily-[7.1^4]dicatena.

II-5.3.3.3.2 *Polycyclic compounds.*

Examples:
1.

1,7-diazy-2,3,4,5,6,8,9,10,11,12,13-undecasulfy-[012.1^{1,7}]dicycle or
1,7-diazyundecasulfy-[012.1^{1,7}]dicycle

Since the compound contains only nitrogen and sulfur, it is not necessary to indicate the locants of all sulfur atoms. Only the locants of the two nitrogen atoms are needed.

2.

3,6-diiodo-1,3,4,6-tetraphosphy-2,5,7-trisulfy-[06.1^{1,4}]dicycle

3.

1-fluoro-2,4,5,7-tetramethyl-3,3,6-trioxo-2,4,5,7-tetraazy-6-carby-1-phosphy-3-sulfy- -[04.3^{1,1}]dicycle

Phosphorus has the locant 1. As the nodal framework comprises only the seven atoms forming the two rings with other groups named as ligands, it is not possible to select the main ring unambiguously by considering only the graph of the molecule. The selection of the main ring is determined by the fact that sulfur is senior to carbon according to the element seniority sequence (see Table II-1).

4.

1,3,3,5,7,7,9,11,11-nonachloro-2,4,6,8,10,12,13-heptaazy-1,3,5,7,9,11-hexaphosphy- -[012.1^{1,5}0^{9,13}]tricycle

5.

1,3,6,8,11,13,16,18-octaboryhexadecasulfy-[020.1^{1,18}1^{3,6}1^{8,11}1^{13,16}]pentacycle

6.

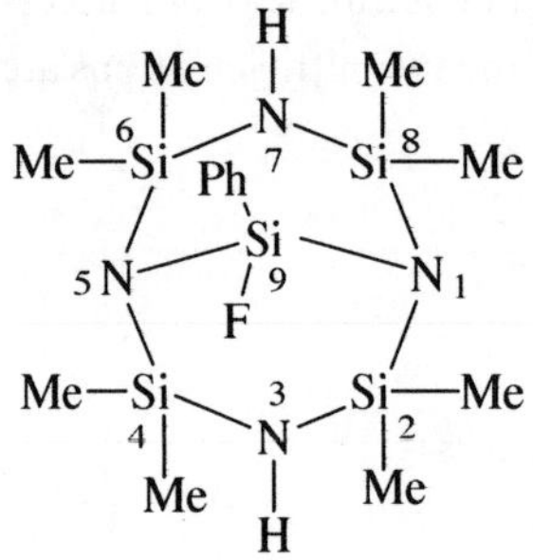

9-fluoro-3,7-dihydrido-2,2,4,4,6,6,8,8-octamethyl-9-phenyl-1,3,5,7-tetraazy-2,4,6,8,9-
-pentasily-[08.1^{1,5}]dicycle

II-5.3.3.3.3 *Assemblies*

Examples:

1.

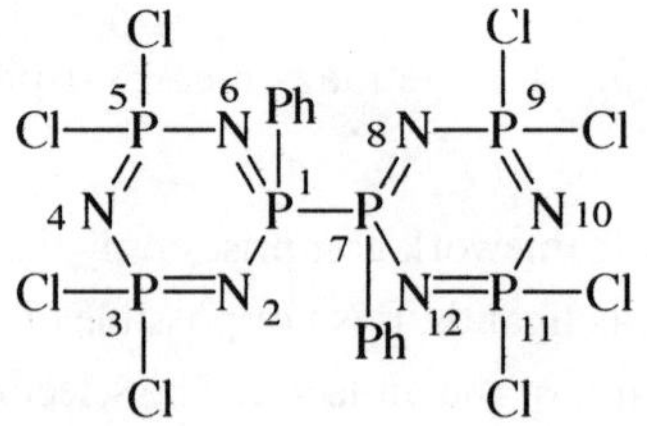

3,3,5,5,9,9,11,11-octachloro-1,7-diphenyl-2,4,6,8,10,12-hexaazy-1,3,5,7,9,11-
-hexaphosphy-[(06)1:7(06)]dicycle

The two modules are identical and thus either can act as a principal module. However, consider a case where P(7) is substituted by As:

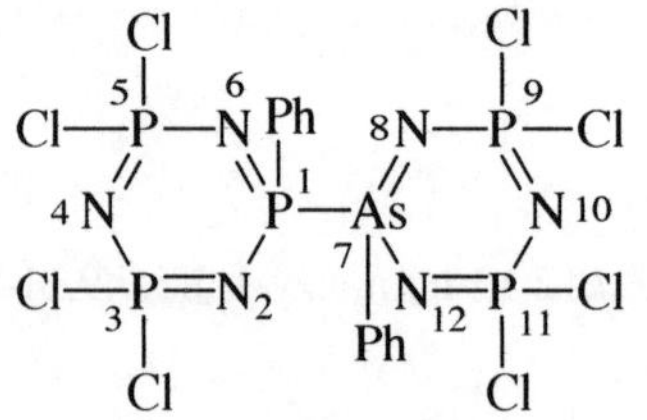

According to general nodal nomenclature (Note 5d) either ring can still constitute the principal module, but as phosphorus is senior to arsenic according to Table II-1, it determines the numbering. The name is therefore 3,3,5,5,9,9,11,11-octachloro-1,7-diphenyl-7-arsy-
-2,4,6,8,10,12-hexaazy-1,3,5,9,11-pentaphosphy-[(06)1:7(06)]dicycle.

2.

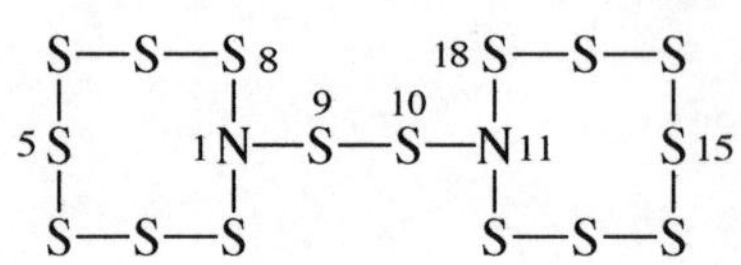

1,11-diazyhexadecasulfy-[(08)1:9(2)10:11(08)]catenadicycle

INORGANIC CHAIN AND RING COMPOUNDS

3.

2,3,4,5,6,8,9,10,11,12-decahydrido-1,3,5,8,10,12-hexaazy-2,4,6,7,9,11-hexabory--[(06)1:7(06)]dicycle

The two modules differ only by their points of attachment. The principal module is determined according to the higher seniority of nitrogen to boron in the element seniority sequence.

II-5.3.3.3.4 *Cage compounds*. The nodal descriptor establishes a connectivity, but not the geometry of the molecule. Most of the inorganic polycyclic compounds form cage structures. If the cage is relatively open and of low symmetry, it is advisable to use the present method involving nodal descriptors.

Example:
1.

This molecule is often regarded as a simple monocyclic compound. Therefore it can be named as 2,4,6,8-tetraazy-1,3,5,7-tetrasulfy-[8]cycle or 2,4,6,8-tetraazy-1,3,5,7-tetrasulfy-[08]cycle.

However, there is significant interaction between the sulfur atoms and therefore in some cases it might be preferable to name the compound as a cage instead of a simple monocycle: 2,4,6,8-tetraazy-1,3,5,7-tetrasulfy-[08.0^{1,5}0^{3,7}]tricycle.

A particularly useful area of application for the nodal descriptor incorporated with the ring and chain nomenclature comprises the low-symmetry borane cages.

Examples:
2.

3,3,7,7,8,8-hexahydrido-1,3,5,7-tetrabory-2,4,6-trihydrony-8-phosphy-[08.0^{1,5}]dicycle

Hydrogen is senior to phosphorus and thus establishes the direction of numbering of the main ring. While bridging hydrogen atoms could be named as bridging ligands, they can conveniently be incorporated in the nodal framework.

3.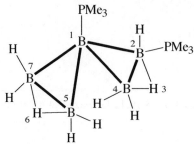

2,4,4,5,5,7,7-heptahydrido-1,2-bis(trimethylphosphane)-1,2,4,5,7-pentabory-3,6-
-dihydrony-[04.3^{1,1}0^{2,4}0^{5,7}]tetracycle

The nodal framework consists of two equivalent rings of four nodes. The main ring is selected by consideration of ligands (PMe$_3$ is senior to H).

4.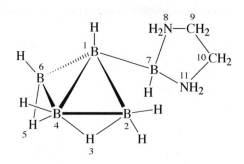

1,2,2,4,6,7,8,8,9,9,10,10,11,11-tetradecahydrido-8,11-diazy-1,2,4,6,7-pentabory-9,10-
-dicarby-3,5-dihydrony-[(06.0^{1,4}0^{2,4}0^{4,6})1:7(05)]pentacycle

If the cage structure is of high symmetry, the description of the structure by the nodal descriptor is more problematic; while it is possible to give a unique name to such a molecule, the visualization of the compound as a symmetric polyhedral cage is not easy.

Examples:
5.

tetraphosphy-[04.0^{1,3}0^{2,4}]tricycle

6.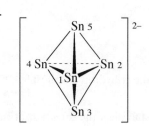

pentastanny-[05.0^{1,3}0^{1,4}0^{2,4}0^{2,5}]pentacyclate(2–)

7.

[Structure: hexatellurium cluster with Te atoms numbered 1-6, charge 4+]

hexatellury-[06.0^{1,3}0^{2,5}0^{4,6}]tetracyclium(4+)

In general, if the structure has a polyhedron of *n* faces, the compound will be named as a polycyclic entity consisting of *n*–1 rings. For most cases, however, high-symmetry cages are best named as clusters.

II-5.3.3.3.5 *Ionic species and ligands.* The names for the ions can be conveniently derived from the corresponding neutral molecules by modifying the names catena and cycle. The charge number of the species is indicated in parentheses at the end of the name.

Examples:

1.

[Structure with S and N atoms, charge 2+]

2,4,7,9-tetraazy-1,3,5,6,8,10-hexasulfy-[010.0^{1,5}0^{6,10}]tricyclium(2+)

2.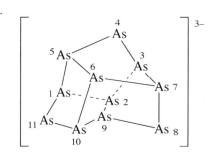

undecaarsy-[011.0^{1,5}0^{2,9}0^{3,7}0^{6,10}]pentacyclate(3–)

3.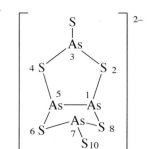

3,7-disulfido-1,3,5,7-tetraarsy-2,4,6,8-tetrasulfy-[08.0^{1,5}]dicyclate(2–)

4.

[structure diagram]

nonadecasulfy-[(07)1:8(5)12:13(07)]catenadicyclium(2+)

When desirable, the location of the charge can be expressed by its nodal locant before the endings -ium or -ate. The charges which are not located in the nodal framework cannot be indicated. Therefore, if the locations of the charges are needed, all pertinent atoms must be included in the nodal framework.

5. Consider the ion in Example 3 of this section. To indicate the location of the charges the nodal skeleton must comprise two acyclic and one polycyclic module: 1,3,5,7-tetraarsy--2,4,6,8,9,10-hexasulfy-[(08.0^{1,5})3:9(1)7:10(1)]dicatenadicycl-9,10-ate(2–). Note that the ending -ate does not imply that the cyclic module carries the charge. The locations of the charges are given unambiguously by the definitive nodal locants.

The construction of the ligand name follows the principles introduced in Chapter I-10 of Note 5b. The names of the cationic and neutral ligands are identical to the names of the free species, but the names of the anionic species are modified by substituting the ending -ate by the ending -ato. The indicated donor atom must be a part of the nodal skeleton.

6.

[structure diagram]

1-oxo-2,4,6,8,9-pentaazy-1,3,5,7-tetrasulfy-[08.1^{1,5}0^{1,5}0^{3,7}]tetracyclato(1–)

When the donor atom O(10) is indicated, the nodal skeleton must consist of two modules: 2,4,6,8,9-pentaazy-10-oxy-1,3,5,7-tetrasulfy-[(08.1^{1,5}0^{1,5}0^{3,7})1:10(1)]catenatetracycl-10-ato.

7.

[structure diagram]

1,2,4,6,8-pentahydrido-1,2,4,6,8-pentabory-3,5,7-trihydrony--[08.0^{1,4}0^{1,6}0^{2,4}0^{4,6}0^{6,8}]hexacycl-2,8-atobis(triphenylphosphane)copper(I)

INORGANIC CHAIN AND RING COMPOUNDS

II-5.3

This can also be named as a coordination compound by considering the borane cage as a ligand.

II-5.3.3.3.6 *Coordination compounds.* In certain cases it is preferable to name the complete coordination entity as a polycyclic compound. This approach is particularly useful for naming polynuclear complexes with purely inorganic chelating or bridging ligands.

Examples:

1.

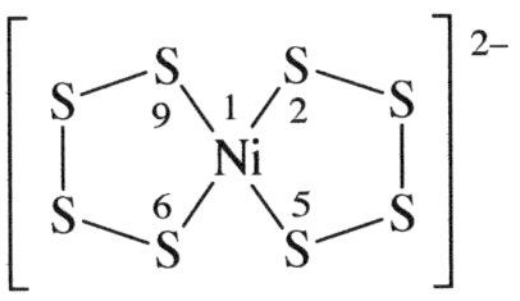

1-nickelyoctasulfy-[05.4^{1,1}]dicyclate(2–)

2.

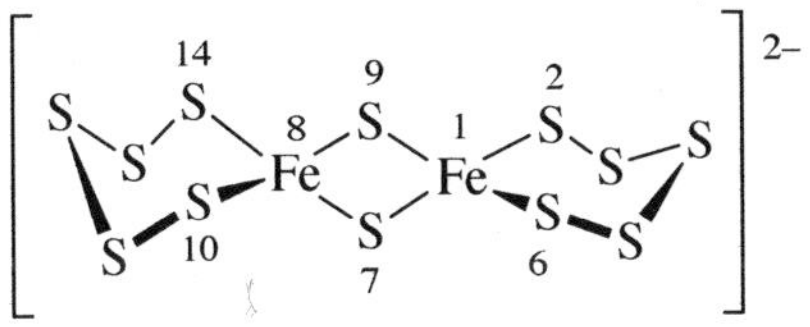

1,8-diferrydodecasulfy-[06.3^{1,1}5^{8,8}]tricyclate(2–)

3.

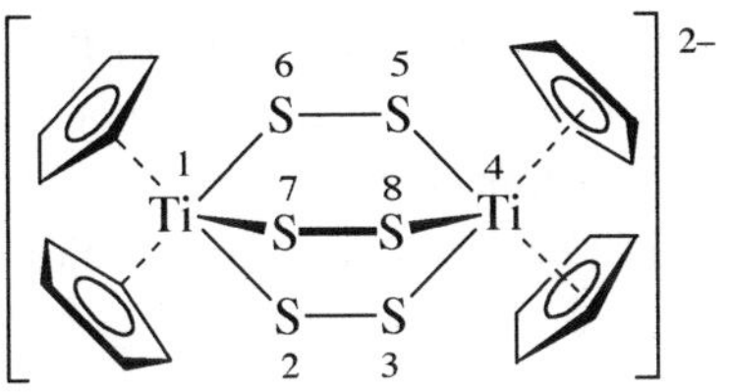

1,1,4,4-tetrakis(cyclopentadienyl)-1,4-dititany-2,3,5,6,7,8-hexasulfy-[06.2^{1,4}]dicyclate(2–)

4.

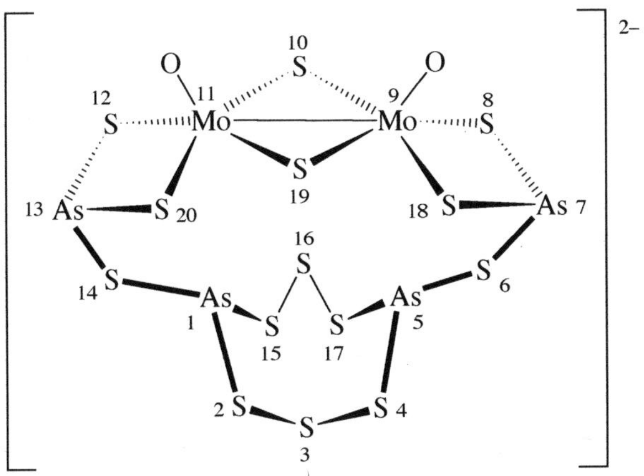

9,11-dioxo-1,5,7,13-tetraarsy-9,11-dimolybdy-tetradecasulfy-
-[014.3^{1,5}1^{7,9}1^{9,11}1^{11,13}0^{9,11}]hexacyclate(2–)

Polynuclear coordination compounds can also be treated as polycyclic compounds. It has to be kept in mind, however, that as in the case of non-metallic cage-compounds the nodal descriptor does not give any indication of the geometry of the cluster; it only defines the connectivity.

Example:
5.

1,1,2,2,2,3,3,5,5-nonacarbonyl-4,6,7-trioxo-4,6,7-tricarby-1,2,3,5-tetracobalty--[06.1^{1,3}0^{1,3}0^{1,5}0^{2,5}0^{3,5}]hexacycle

II-5.4 CONCLUSION

This chapter presents a new additive method to name inorganic chain and ring compounds. Though it can be applied to any molecular species, its use is mainly intended for compounds which are mainly or exclusively made from atoms other than carbon. The application of the nodal descriptor enables the naming of complicated polycyclic and branched chain compounds as well as compounds consisting of cyclic and acyclic modules. The selection of the nodal skeleton renders it possible to compare structurally analogous species regardless of their chemical identity.

The method is particularly suitable for low symmetry ring and cage compounds and whenever it is important to establish the connectivity between the atoms. For high symmetry cages and polynuclear coordination compounds the use of more exact structural descriptors is needed.

II-6 Graphite Intercalation Compounds

CONTENTS

II-6.1 General considerations
II-6.2 Vocabulary
  II-6.2.1 The reaction
  II-6.2.2 The intercalated layer
  II-6.2.3 The carbon layers
  II-6.2.4 Stages
II-6.3 Classification of graphite intercalation compounds
  II-6.3.1 Donor and acceptor compounds
  II-6.3.2 Graphite salts
  II-6.3.3 Binary, ternary, quaternary, *etc.*, compounds
  II-6.3.4 Multi-intercalation compounds
  II-6.3.5 Heterostructures
II-6.4 Non-formula-based description of graphite intercalation compounds
II-6.5 Formulation of graphite intercalation compounds
II-6.6 Structural notation
  II-6.6.1 Description of stacking sequence
  II-6.6.2 Description of the crystal lattice
  II-6.6.3 Description of the epitaxial layer
  II-6.6.4 Symbols for various distances

II-6.1 GENERAL CONSIDERATIONS

In graphite intercalation compounds there are many varied and complex structures. The following recommendations are designed to cover the majority of cases (Note 6a).

Several types of compounds are derived from graphite. One may distinguish compounds which have species bound by covalent two-electron bonds to the carbon atoms, and compounds in which there is a charge transfer to or from the graphene layers (see Section II-6.3) of the graphite structure. The species (atoms, ions or molecules) are, in both cases, inserted between the carbon sheets. This results in a more or less pronounced expansion which may be detected by X-ray diffraction and which occurs in a direction perpendicular to the carbon sheets.

Necessarily, the carbon layers lose their planarity in the case of covalent bonding and convert to a puckered-layer structure. Covalent bonds are formed by reactions of graphite with elements or groups of high electronegativity, *e.g.* fluorine (which yields graphite

---

Note 6a. See also A.D. McNaught and A. Wilkinson, *Compendium of Chemical Terminology*, Blackwell Scientific Publications, Oxford, 1997; Recommended Terminology for the Description of Carbon as a Solid, *Pure Appl. Chem.*, **67**, 473 (1995).

fluorides), oxygen or hydroxyl groups (which give graphite oxide, also called graphitic acid), *etc*. The disruption of the π-electron density generally leads to a drastic decrease of electrical conductivity.

In the case of a charge transfer interaction, the carbon sheets retain their planar structure (apart from the possible presence of local defects) even though the guest material is found in the expanded interlayer intervals of the host structure. The compounds thus formed are called *graphite intercalation compounds*. Electronic interaction with the carbon sheets of the graphite host structure leads, in general, to a remarkable increase of in-plane electrical conductivity, although this may be offset by the presence of defects originally present or formed during the intercalation process. Even though this is not sufficiently recognized in the literature, it should be kept in mind that, in general, these graphite intercalation compounds have a more or less broad range of compositions and are *non-stoichiometric compounds*.

The term graphite intercalation compound(s) has been abbreviated to GIC. In order to avoid confusion in computer-aided searches, it is recommended that this abbreviation be avoided in titles since IC generally stands for integrated circuit.

II-6.2 VOCABULARY

II-6.2.1 The reaction

The term *intercalation* is used to describe the overall process whereby a compound is formed without loss of planarity for the layers of carbon atoms, with heteroatoms or molecules residing between the original layers of carbon, *etc.*, as described in Section II-6.1 above. The reverse process, *i.e.* the loss of some or all of the intercalated species, by whatever means, is *de-intercalation* or *disintercalation*.

II-6.2.2 The intercalated layer

In graphite intercalation chemistry, the term intercalated must refer to the guest (*e.g.* potassium or bromine) and not the host species (graphite). Consequently, the description intercalated potassium (or bromine) in graphite is correct, but not intercalated graphite. The present participle intercalating should not be used to describe a state but rather a process, hence intercalating potassium describes those atoms which are reacting with graphite to become finally intercalated potassium. To prevent possible ambiguities, the substantive intercalant should be avoided. However, by analogy with an IUPAC recommendation (Note 6b), intercalate can be used to describe the guest intercalated in the host lattice.

II-6.2.3 The carbon layers

The term graphite designates a mineral as well as the crystal structure of an allotropic form of elemental carbon. In graphite, planar sheets of carbon atoms, with each atom bound to three neighbours in a non-compact honeycomb structure, are stacked regularly with three-

---

Note 6b. Compare with 'adsorbate': 'Definitions, Terminology and Symbols in Colloid and Surface Chemistry', *Pure Appl. Chem.*, **31**, 577 (1972), Section 1.1.4; **54**, 2201 (1982), Section 1.2.1.

dimensional order (Note 6c). Graphitic carbon is only applicable to materials which give rise to at least a modulation of the *hk* reflections in X-ray diffraction. It is therefore not correct to speak of graphite layers when meaning single, two-dimensional carbon sheets. Even the terms carbon layer or carbon sheet are inappropriate.

The suffix -ene is used for fused polycyclic aromatic hydrocarbons, even when the root of the name is of trivial origin, *e.g.* naphthalene, anthracene, coronene, *etc*. A single carbon layer of the graphitic structure can be considered as the final member of this series and the term graphene should therefore be used to designate the individual carbon layers in graphite intercalation compounds.

II-6.2.4 Stages

A peculiarity of graphite intercalation compounds is their tendency to form regularly stacked structures in which the intercalate is only found in a fraction of the interlayer space between the graphene layers. In a first stage (or stage 1, st.1) compound, single layers of graphene alternate regularly with single layers of intercalated species (the latter may be more than one atom thick). In a second (st.2), third (st.3), *etc.*, stage compound, two, three, *etc.*, graphene layers separate two successive layers of intercalate.

```
OOOOOO      OOOOOO      ←— intercalated layers —→      OOOOOO
------      ------                                     ------
OOOOOO      ------      ←— graphene layers —→          ------
            OOOOOO                                     ------
                                                       OOOOOO

stage 1     stage 2                                    stage 3
```

According to accepted rules (Note 6d), the unit forming a polymer by continuous succession is called a constitutional repeating unit (CRU) (see also Chapter II-7). By analogy, the layer sequence of a stage, as shown above, would be the CRU of a graphite intercalation compound.

In another structural model, second, third, *etc.* stage compounds are characterized by the fact that the area over which the layer of intercalated species is found between each pair of successive graphene layers is one half, one third, *etc.* of the interlayer area (Note 6e).

In contrast to intercalation compounds given by other species such as TaS_2, high number stages (10, 11, *etc.*) have been observed with graphite intercalation compounds, as well as compounds with fractional stage number (in which, for instance and with reference to the first structural model described above, two out of three interlayer spaces contain the intercalate) or with average stage numbers. In such cases, a full description and explanation should be given in the text. A clear distinction should be made between these cases and occasional stacking faults.

Note 6c. See International Committee for Characterisation and Terminology of Carbon: First Publication of 30 Tentative Definitions, Terms No.3 and 6; *Carbon*, **20**, 445 (1982).
Note 6d. Basic Definitions of Terms Relating to Polymers, 1974, *Pure Appl. Chem.*, **40**, 477 (1974).
Note 6e. A. Hérold, *Crystallochemistry of Carbon Intercalation Compounds*, in *Intercalated Layered Materials* (F. Lévy, ed.), D. Reidel, Dordrecht, 1979, pp. 323–421 (in particular p. 400).

II-6.3 CLASSIFICATION OF GRAPHITE INTERCALATION COMPOUNDS

II-6.3.1 Donor and acceptor compounds

In graphite intercalation compounds, the graphene layers either accept electrons from or donate electrons to the intercalated species. It has now been generally accepted, however, that the description of the electronic exchange in these compounds should be specified from the standpoint of the acceptor or donor properties of the intercalated species rather than from that of the graphene layers, as customary with doped semi-conductors. Potassium–graphite is therefore a donor graphite intercalation compound, whereas bromine–graphite, or arsenic pentafluoride–graphite, or graphite hydrogensulfate are acceptor compounds.

For the graphite intercalation compounds in which different intercalated species located in specific regions of the interlayer space bear charges of opposite sign, the description mixed donor-acceptor intercalation compound can be used.

The extent of charge transfer has frequently been defined by the fraction of intercalated species carrying a charge. This practice is discouraged in view of the non-stoichiometric nature of the graphite intercalation compounds. Also, in many cases the layers of intercalate are known to differ in order and density. It is therefore recommended that the charge transfer be defined by the average charge accepted or donated per carbon atom of the graphene layer. However, it must be borne in mind that in stages higher than the second, most of the charge may be concentrated on the graphene layers immediately adjacent to the intercalated layers. Authors should therefore specify how the value of the charge transfer is defined.

II-6.3.2 Graphite salts

Following long-established usage, the name graphite salt can be given to a subclass of acceptor compounds formulated $C_m^+X^-\cdot n\ HX$, in which HX represents a molecule of Brønsted acid. This name reflects the fact that these compounds are known to contain discrete ions, such as HSO_4^- or NO_3^-, as evident from the method of preparation which involves intercalation of the graphene layers by chemical or electrochemical means in the presence of the acid. The space between the anions in the intercalated layers is filled by molecules of the acid.

Similarly, compounds containing intercalated cationic layers, with the graphene layers negatively charged, have been called graphitides, in accordance with the recommendations applicable to names of salts of non-complex anions. In view of the arguments developed in Section II-6.2.3, the name graphenide may be more appropriate.

II-6.3.3 Binary, ternary, quaternary, *etc.* compounds

The order of a graphite intercalation compound indicates the number of components in it: binary compounds are those which contain a single chemical species besides the carbon in the original graphite, *i.e.* in which the total number of species is two. In this connection, ions cognate to an intercalated neutral atom or molecule and co-intercalated with it (as in the graphite salts of Section II-6.3.2) will not be considered to be a different chemical species. Ternary, quaternary, *etc.* graphite intercalation compounds respectively contain two, three, *etc.* different chemical species besides the original carbon of the graphene.

II-6.3.4 Multi-intercalation compounds

The term bi-intercalation compound describes a ternary graphite intercalation compound in which two different guests occupy separate interlamellar species of the host structure.

For example:

```
OOOOOO
------
xxxxxxxxx
------
OOOOOO
```

This nomenclature applies, with appropriate modifications, to compounds in which three, four, *etc.* different guests occupy separate and successive interlayer spaces of the host structure, forming tri-intercalation, tetra-intercalation, *etc.* compounds. These terms do not apply to solid solutions of one guest species in the intercalated layers of another guest species.

II-6.3.5 Heterostructures

Some attempts at preparing ternary compounds result in the formation of two phases, shown by the presence of X-ray diffraction patterns for both. Since the individual domains are of such microscopic size that the mechanical separation of the two phases is excluded, these mixtures can be described as heterostructures.

II-6.4 NON-FORMULA-BASED DESCRIPTION OF GRAPHITE INTERCALATION COMPOUNDS

The name of a graphite intercalation compound is formed by combining the noun graphite with the name of the intercalated substance, separated by a long dash ('en-dash'). If necessary, the composition and the stage may be added, between parentheses. The oxidation number of an element may be specified. In a non-formula-based description, the custom is to put the intercalated guest species first, whether it be donor or acceptor. In some cases (especially with graphite salts), it is recommended that the noun graphite is put first (*vide infra*), with no hyphen between it and the name of the anion.

Examples:
1. lithium–graphite (1/6, st.1)
2. potassium–graphite (1/24, st.2)
3. iodine monochloride–graphite ($\approx$1/8, st.2)
4. iron(III) chloride–graphite ($\approx$1/18)
5. graphite tetrachloroaurate(III)

Compounds may exhibit a considerable range of compositions and the exact constitution should only be given when the sample under discussion has been analysed. Similarly, doubts as to the exact value of the stage number preclude it from being given.

In ternary compounds containing co-intercalated molecules which solvate the anion or cation between the layers of graphene, the names of the molecules are always placed after that of the solvated species.

Examples:
6. potassium–benzene–graphite (1/2, 2/24, st.1)
7. graphite tetrafluoroborate–tetrahydrofuran

II-6.5 FORMULATION OF GRAPHITE INTERCALATION COMPOUNDS

Formulae should be specified only if accurately known, otherwise it is preferable to use a non-formula-based description such as potassium–graphite (st.2). No stage number should be given if it is undefined.

In agreement with the recommendations of Section I-4.6.1.2 (Note 6f), the electropositive constituent must be placed first in the formula.

Examples:
1. SrC_6 (st.1)   strontium–graphite (1/6, st.1)
2. $FeC_{14}Cl_3$ (st.2)   iron(III) chloride–graphite (1/14, st.2)

In ternary or higher order compounds, similar constituents appear in the formula in alphabetical order of the chemical symbol.

Example:
3. $Cs_{1-x}K_xC_8$ (st.1)   caesium–potassium–graphite (st.1)

With solvated ions, the symbol of the solvating molecules follows that of the ion.

Examples:
4. $Ba(NH_3)_{2.5}C_{10.9}$ (st.1)   barium–ammonia–graphite (1/2.5/10.9 st.1)
5. $C_{24}{}^+[PF_6]^-(thf)_n$ graphite–hexafluorophosphate–tetrahydrofuran

It is also possible to specify the co-intercalated molecules after a centre dot.

Example:
6. $C_m{}^+[PF_6]^- \cdot n$thf

Bi-intercalation (*i.e.* different species intercalated in different interlayer spaces) should be made apparent in the formula.

Examples:
7. $TlC_{12.5}Br_{3.2} \cdot TlC_{12.5}Cl_{3.6}$ (st.1)   thallium bromide–thallium chloride–graphite (st.1)

Note 6f. *Nomenclature of Inorganic Chemistry, Recommendations 1990*, Blackwell Scientific Publications, Oxford, 1990.

8. $CsC_8 \cdot K_2H_{4/3}C_8$ (st.1)   caesium–hydrogen–potassium–graphite (st.1)

It is desirable that any non-stoichiometry of the intercalation compounds is also expressed in the formulation. Thus, $K_{1-x}C_8$ implies that some potassium atoms are missing in the two-dimensional intercalated layers of KC_8. A commensurate structure necessarily implies an integral number of carbon atoms and if, for example, $x = 0.1$ in $K_{1-x}C_8$, it would be wrong to give the formula as $KC_{8.9}$. Similarly, the formula $NiC_{13.2}Cl_{2+x}$ implies (i) that there is an excess of chlorine with respect to the integral stoichiometry Cl/Ni = 2, and (ii) that no simple integral ratio exists for the number of C atoms to each Ni atom in nickel(II) chloride–graphite. Hence the intercalated $NiCl_2$ layers are non-commensurate with respect to the layers of graphene.

However, it should be recognized that it is not always possible to indicate unambiguously by a chemical formula whether a given structure is commensurate or non-commensurate with the graphene layers. Moreover, an ambiguity arises from the use of non-integral indices in the formulae. Whereas the index 0.8 in $Li_{0.8}C_6$ is meant to indicate that only 80% of the Li sites in the first-stage lithium graphite derivative are occupied, the same index in $K(furan)_{0.8}C_8$ gives the actual maximum content of furan molecules since these occupy *all* the available space, and not just 80% of it. If the appropriate information is available, the type of defect responsible for the non-integral stoichiometry can be indicated using standard notation, *i.e.* V for vacancies and i for interstitials (Section I-6.4.2 of Note 6e).

II-6.6 STRUCTURAL NOTATION

II-6.6.1 Description of the stacking sequence

In the hexagonal form of graphite, the graphene layers are stacked so that all even-numbered layers are shifted with respect to the odd-numbered layers by one-third of the crystallographic parameter, as reflected in the notation ABAB ... Similarly, graphene layers in different relative positions can be indicated by capital Roman letters (A, B, C ...). For the intercalated layers, a distinction should be made between the following two cases:

- for commensurate layers in epitaxy on the graphene layers, lower case Greek letters (α, β, γ, ...) can be used to distinguish between individual layers;
- non-commensurate or disordered layers can be identified by χ (chi) or ξ (xi).

Thus, the structure of LiC_6 (st.1) is described by $C\alpha C\alpha C\alpha$.., while KC_8 (st.1), more correctly formulated as $(KC_8)_4$ (st.1), is described by $C\alpha C\beta C\gamma C\delta$ ....

II-6.6.2 Description of the crystal lattice

The symmetry of the crystal lattice of a given stage may be indicated by the modified Gard system (Note 6g) (already used for other polytypes), with an Arabic number and a capital Roman letter preceding the chemical formula. The numerical symbol indicates the number of constitutional repeating units (CRU) in the identity period, and the letter indicates the symmetry (H = hexagonal, T = trigonal with hexagonal Bravais lattice, R = trigonal with rhombohedral Bravais lattice, O = orthorhombic).

Examples:
1. $1H$–LiC_6 (st.1)   stacking sequence: $A\alpha A\alpha A\ldots$
2. $2H$–KC_{24} (st.2)   stacking sequence: $AB\chi BA\chi\ldots$
3. $3R$–CsC_8 (st.1)   stacking sequences: $A\alpha A\beta A\gamma\ldots$
4. $4O$–KC_8 (st.1)   stacking sequence: $A\alpha A\beta A\gamma A\delta\ldots$

II-6.6.3 Description of the epitaxial layer

The ordered relationship between graphene layers and intercalated layers in the (a,b) planes is indicated by the notation for epitaxial overlayers (Note 6h). The size of the new unit cell, normalized with respect to the a and b parameters of the graphene layers, is given in parentheses, and rotation with respect to the graphite unit cell is indicated by $R\,\theta°$.

Examples:
1. LiC_6 (st.1); $(\sqrt{3}\times\sqrt{3})\,R\,30°$
2. KC_8; (2×2)
3. CsC_{28} (st.2); $(\sqrt{7}\times\sqrt{7})\,R\,19.1°$

II-6.6.4 Symbols for various distances

The symbol I_c, often used in the literature to indicate the thickness of the CRU (*i.e.* the distance between two successive intercalated layers), can be confused with I, which stands for the body-centred unit cell. To conform with the accepted crystallographic symbolism, the following symbols should be used:

- d_G is the distance between two successive carbon layers in graphite, hence $d_G \approx 335$ pm;
- d_i is the distance between two successive layers of intercalate; this is not necessarily the crystallographic identity period c, which will be a multiple of d_i if screw axes or translation vectors are present; thus, in the case of $(KC_8)_4$, $c \approx 2160$ pm $= 4\times d_i$;
- d_g is the distance between two adjacent layers of graphene if this has been found to be different from 335 pm;
- d_1 is the repeat distance of the first stage compound with the same layer of intercalate as the one in the intercalation compound under consideration, irrespective of whether this first stage compound can exist or not;

Note 6g. Nomenclature of Polytype Structures. Report of the International Union of Crystallography *ad hoc* committee on the nomenclature of disordered, modulated and polytype structures, *Acta Cryst. Sect. A*, **A40**, 399 (1984).
Note 6h. E.A. Wood, *J. Appl. Phys.* **35**, 1306 (1964); R.J. Estrup and M.G. McRae, *Surf. Sci.*, **25**, 1 (1972) (in particular p. 33).

- Δd is the apparent thickness of the intercalated layer, given by $\Delta d = d_I - d_G$ or $\Delta d = d_I - d_g$.

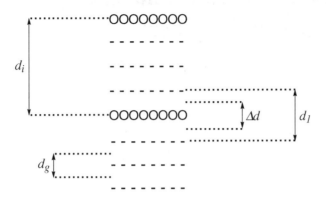

- d_∞ is the in-plane C–C distance in a layer of graphene; it is not necessarily the same distance as in graphite (≈ 142.1 pm) owing to the charge transfer caused by intercalation;
- $\Delta d_\infty = u = d_\infty - 142.1$ pm is the variation of in-plane C–C distance due to the charge transfer.

II-7 Regular Single-Strand and Quasi-Single-Strand Inorganic and Coordination Polymers

CONTENTS

II-7.1 Introduction
II-7.2 Fundamental principles
    II-7.2.1 Identification of the constitutional repeating unit (CRU)
    II-7.2.2 Orientation of CRU
    II-7.2.3 Naming the CRU
II-7.3 Recommendations
    II-7.3.1 The general polymer name
        II-7.3.1.1 The use of poly and square brackets
        II-7.3.1.2 Numerical prefixes
        II-7.3.1.3 Use of the term catena
        II-7.3.1.4 Specification of end groups
    II-7.3.2 Seniority rules for selection of a preferred constitutional repeating unit
        II-7.3.2.1 Choice of the senior constituent subunit
        II-7.3.2.2 Choice of the preferred direction along the polymer chain for the sequential citation of the constituent subunits of the CRU
    II-7.3.3 Regular single-strand inorganic and coordination polymers
        II-7.3.3.1 CRUs with homoatomic backbones
        II-7.3.3.2 CRUs with backbones consisting of one central atom and one bridging ligand
        II-7.3.3.3 CRUs consisting of more than one central atom and no more than one bridging ligand between each central atom of the polymer
    II-7.3.4 Regular quasi-single-strand coordination polymers
        II-7.3.4.1 CRUs with backbones consisting of one central atom and two or more bridging ligands or chelating ligands
        II-7.3.4.2 CRUs containing more than one central atom
    II-7.3.5 Single-strand and quasi-single-strand coordination polymers with polynuclear coordination centres
    II-7.3.6 Regular single-strand and quasi-single-strand inorganic and co-ordination polymers with ionic CRUs
    II-7.3.7 Stereochemical configuration of a CRU
    II-7.3.8 End groups of linear inorganic or coordination polymers
        II-7.3.8.1 Use of the terms α and ω
        II-7.3.8.2 Choice of end groups
        II-7.3.8.3 Ionic end groups

II-7.1 INTRODUCTION

The previously described system for naming polymers in terms of their structure (Notes 7a and 7b) has been concerned with linear organic polymers, primarily those defined as regular single-strand polymers (Note 7b), and has followed as closely as possible established principles of organic nomenclature (Note 7c). Accordingly, constituent subunits of the smallest repeating structural unit, named as bivalent substituent groups, are combined additively to form the name of the constitutional repeating unit. Extension of this method to linear inorganic and/or coordination polymers is seriously limited by the general lack of a system for naming bivalent substituent groups, partly because of the basic difference in philosophy between inorganic and organic nomenclature systems, and partly because the constituent units of the constitutional repeating unit in most inorganic and coordination polymers are not bivalent substituent groups in the usual sense.

The system presented here is based on previous recommendations (Note 7d) and is designed to name, uniquely and unambiguously, regular inorganic and/or coordination linear polymers. The constituent subunits are formulated according to the usual chemical principles of covalent and/or coordinate bonding, and the structures are described by a constitutional repeating unit with at least one terminal constituent subunit that is connected through only one atom to the other identical constitutional repeating units, or to an end group. Ladder structures are thus excluded.

A regular linear polymer that can be described by a preferred constitutional repeating unit in which both terminal constituent subunits are connected through single atoms to the other identical constitutional repeating units or to an end group is called a regular single-strand polymer (see Note 7b).

Example:

1.

single atoms

A regular linear polymer that can be described by a preferred constitutional repeating unit in which only one terminal constituent subunit is connected through a single atom to the other identical constitutional repeating units or to an end group is a quasi-single-strand polymer, *i.e.* it does not fit the definition of regular single-strand polymers (Note 7b), but can be named in the same manner.

Note 7a. *Macromolecules*, **1**, 193 (1968).
Note 7b. IUPAC Information Bulletin Appendices on Tentative Nomenclature, Symbols, Units and Standards, No. 29, Nov., 1972; *Macromolecules*, **6**, 149 (1973); *Pure Appl. Chem.*, **48**, 373 (1976).
Note 7c. *Nomenclature of Organic Chemistry*, Pergamon Press, Oxford, 1979; (a) Recommendation D-6.22, pp. 411–412; (b) Recommendation A-1.1, p. 5.
Note 7d. *Nomenclature of Inorganic Chemistry, Recommendations 1990*, Blackwell Scientific Publications, Oxford, 1990; Section I-10.8.4. See also W. V. Metanomski in *Compendium of Macromolecular Nomenclature*, 1st Edition, Blackwell Scientific Publications, Oxford, 1991.

Example:

2.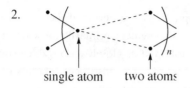

        single atom   two atoms

Established principles of inorganic and coordination nomenclature (Note 7d) are used as far as is consistent with the definitions (Note 7e) and basic principles (Notes 7a and 7b) of polymer nomenclature already published. As in the nomenclature of organic polymers, these recommendations apply to structures, not substances, that may be idealized representations of complex systems. Polymeric substances usually include a number of different structures, and a complete description of a polymer molecule would have to include such items as degree of steric regularity, chain imperfections, random branching, *etc.*, resulting in extremely complex nomenclature. In any event, it is useful to consider a polymer in terms of a single structure that may itself be idealized. To the extent that an inorganic or coordination polymer can be represented as a linear combination of regularly repeating constitutional (structural) units, it can be named according to the following recommendations. End goups may be included, if desired.

A few polymers with inorganic backbones in which the bonding is primarily covalent have trivial or semi-systematic names of long standing, *e.g.* poly(dimethylsiloxane) for $\{-Si(CH_3)_2-O-\}_n$ (see Note 7f) and poly(dichlorophosphazene) for $\{-PCl_2=N-\}_n$ (see Note 7g); some of these polymers can also be named by the principles for naming organic polymers, *i.e.* by using bivalent substituent groups (Notes 7a and 7b), *e.g.* poly[oxy(dimethylsi-lylene)] for $\{-Si(CH_3)_2-O-\}_n$, and poly[nitrile(dichlorophosphoranylidyne)] for $\{-PCl_2=N-\}_n$ (see Note 1b, p. 10). There is no objection to the use of names based on the principles for naming organic polymers, if the names for the bivalent substituent groups in the structure are clearly established. However, for some structures the use of the recommendations given below provides unambiguous names with much less artificiality.

II-7.2 FUNDAMENTAL PRINCIPLES

The system of nomenclature for regular single-strand and quasi-single-strand inorganic and coordination polymers presented here is governed by the same fundamental principles of polymer nomenclature developed for single-strand organic polymers (Notes 7b and 7h). It is

---

Note 7e. *Pure Appl. Chem.*, **40**, 479 (1974): (a) term 3.3, p. 482.

Note 7f. Names of siloxane polymers have been based on a $-SiH_2-O-$ repeating unit called siloxane which, together with substituent names, is enclosed in parentheses or brackets and preceded by the prefix poly. Low molecular weight siloxane polymers can be named by using the prefix oligo or a numerical prefix in place of poly as described in the organic polymer recommendations (Notes 7a and 7b). On the other hand, names of specific low molecular weight acyclic siloxanes with the general formula $H_3Si[-O-SiH_2-]_nO-SiH_3$ are formed according to Recommendation D-6.22 (Note 7c). This results in similar but not identical names. For example when $n = 2$ in the general formula above, the name would be tetrasiloxane. In these names, substituents are cited as prefixes to such parent names, for example decamethyltetrasiloxane.

Note 7g. This name is a hybrid of additive and substitutive nomenclature that does not satisfy the recommendations of either system.

Note 7h. See also A.D. McNaught and A. Wilkinson, *Compendium of Chemical Terminology*, Blackwell Scientific Publications, Oxford, 1997 and Glossary of Basic Terms in Polymer Science, *Pure Appl. Chem.*, **68**, 2287 (1996).

based on the selection and naming of a constitutional repeating unit (CRU), defined (Note 7e) as the smallest structural unit the repetition of which describes the polymer structure. The name of the polymer is the name of this repeating unit prefixed by the terms poly, *catena*, or other structural indicator, and designations for end groups, if desired.

The name of the CRU is formed by citing, in order of appearance along the chain of the CRU, the names of its constituent subunits, which are the longest structural fragments of the CRU that can be named by the established principles of inorganic and/or coordination nomenclature (Note 7i). Accordingly, bridging ligands are not broken into subunits smaller than those named by principles of coordination nomenclature for ligands (Note 7d).

Although this procedure will result in an unambiguous name, it will not necessarily give a unique name. To obtain a unique name, a single, preferred CRU must be selected. This may be accomplished by the following procedure:

(a) identify the CRU and its constituent subunits;
(b) orient the CRU, *i.e.* determine the subunit that will begin the citation of the CRU and the direction to proceed along the backbone of the polymer chain in writing the rest of the CRU from left to right;
(c) name the CRU in two basic steps:
   (i) name the subunits by established inorganic and/or coordination nomenclature principles;
   (ii) assemble the names of the subunits according to the preferred direction of citation to form the name of the CRU.

(It is important to identify and orient the CRU as far as possible before assembling the name of the CRU. However, occasionally the choice of a unique orientation will depend on names for individual subunits).

II-7.2.1 Identification of the constitutional repeating unit (CRU)

In many cases, the polymer structure is simple enough so that the CRU and its constituent subunits can readily be identified.

Example:

In the polymer $\left(\begin{array}{c}F\\|\\-Au-F-\\|\\F\end{array}\right)_n$, the two possible CRUs are $-\overset{F}{\underset{F}{\overset{|}{Au}}}-F-$ and $-F-\overset{F}{\underset{F}{\overset{|}{Au}}}-$.

Note 7i. In the rules for naming organic polymers (Notes 7a and 7b), the constituent subunits of the CRU are the largest structural fragments than can be named as bivalent or multivalent substituent groups according to the established principles of organic nomenclature (Note 7c). Single central atoms, mononuclear coordination centres, and bridging ligands are constituent subunits of CRUs in regular single-strand and quasi-single-strand inorganic and coordination polymers. Polynuclear coordination centres are used as subunits only under certain conditions (see discussion in Section II-7.2.3).

POLYMERS

However, in more complex cases and occasionally in some simple cases it may be necessary to draw a fairly long segment of the polymer chain in order to identify the possible constitutional repeating units.

Examples:

1. In the polymer $-(-Ag-CN-Ag-CN-)_n-$, the possible constitutional repeating units are: $-Ag-CN-$; $-Ag-NC-$; $-CN-Ag-$; and $-NC-Ag-$ (as already stated the bridging ligand $-CN-$ is not broken up into smaller subunits).

2. In the polymer

$$\left(-Hg-S \overset{NEt_2}{\underset{Br}{\hat{C}}} O-Hg \overset{Br}{\underset{Br}{|}} -O \overset{NEt_2}{\hat{C}} S-Hg-O \overset{NEt_2}{\hat{C}} S-Hg \overset{Br}{\underset{Br}{|}} -S \overset{NEt_2}{\hat{C}} O- \right)_n$$

the possible constitutional repeating units are:

[structures of four possible constitutional repeating units shown, each being a cyclic permutation of the segment above]

II-7.2.2 Orientation of the CRU

The constituent subunit of the CRU at which the citation of the CRU begins is the central atom (or coordination centre) of highest seniority, *i.e.* the most preferred central atom according to the set of hierarchical rules given in Section II-7.3.2 below. The centre is normally written as the left terminal subunit of the CRU.

The preferred direction along the polymer chain from the senior subunit for the sequential citation (from left to right) of the other constituent subunits in the CRU is governed by three major factors considered in order until a definitive decision is reached.

(a) A single-strand CRU is preferred to a quasi-single-strand CRU, *i.e.* a CRU with both terminal constituent subunits connected to other identical constitutional repeating units or to an end group through single atoms is preferred to a CRU with only one terminal constituent subunit connected to other constitutional repeating units or to an end group through a single atom.

(b) The preferred direction is defined by the shortest path, measured in terms of the number of atoms, in the polymer backbone from the senior subunit to a subunit of equal seniority, or to a subunit next in seniority.

(c) When all paths between the senior subunit and a subunit of equal seniority, or a subunit ranking next in seniority, are of equal length, the preferred direction is along the path that includes constituent subunits of higher seniority. The paths between subunits of equal seniority or between the senior subunit and the subunit next in seniority necessarily involve subunits of lesser seniority, and often will consist of organic ligands. Hence, the hierarchical order of subunits prescribed from linear organic polymers (Notes 7a and 7b) may be needed to determine the preferred direction.

Further refinements to the general principles are given in Section II-7.3.2, and are illustrated in subsequent sections dealing with the naming of specific polymers.

II-7.2.3 Naming the CRU

The name of a CRU of a single-strand and quasi-single-strand inorganic or coordination polymer is based on a backbone consisting of central atoms and bridging ligands where present. All inorganic or coordination polymers have one or more central atoms, but may or may not have bridging ligands. Homoatomic inorganic polymers are considered to consist of central atoms only. Coordination centres, mononuclear or polynuclear, and their associated ligands, except for ligands between central atoms in the backbone, if any, are named by the usual principles of coordination nomenclature. Bridging ligands are named as ligands prefixed by the Greek letter μ.

Selection of the largest structural fragments in the backbone that can be assigned multivalent substituent group names as subunits of a CRU is a fundamental principle in naming linear organic polymers. For naming inorganic and coordination polymers, this principle is applied to the selection of bridging ligands in the CRU. When there is a choice, the largest group that can be named by the accepted methods for naming polydentate ligands is chosen. For example, in the polymer shown below, the CRU could be considered as two central atoms connected by sulfur ligands.

However, the principle of largest bridging ligand requires the bridging ligand to be tetrathiophosphato(3−). Strict application of this principle to inorganic or coordination polymers would lead to the selection of polynuclear coordination centres as the largest structural fragment in the backbone. Since there are no officially accepted rules for uniquely naming and/or numbering certain types of polynuclear coordination centres, it is not yet convenient in some cases to use polynuclear coordination centres as subunits of a CRU in inorganic and coordination polymers. Hence, the principle of largest subunit is not always applied to coordination centres of a CRU and in this set of recommendations polynuclear coordination centres are used as subunits of CRUs only when it is *not* convenient to express such structural units in terms of their mononuclear coordination centres (see Section II-7.3.5). However, for illustrative purposes, names using polynuclear subunits are given as alternatives for some of the examples in the recommendations that follow.

Once the names of the constituent subunits of the CRU are determined, the CRU name is formed by citing the name of the senior subunit followed by the names of the other constituent subunits as they occur in the preferred direction along the polymer chain.

II-7.3 RECOMMENDATIONS

II-7.3.1 The general polymer name

II-7.3.1.1 *The use of poly and square brackets*

A name of a polymer for which the constitutional repeating unit is known, but with a dimensional structure (Note 7j) that may not be known or that need not be specified, consists of the prefix poly followed by the names of the contitutional repeating unit (see Sections II-7.3.3–II-7.3.7) enclosed in square brackets, *e.g.* poly[CRU] (Note 7k).

II-7.3.1.2 *Numerical prefixes*

If it is desired to specify the number of constitutional repeating units, the appropriate numerical prefix (Note 7c) may be used in place of the prefix poly, *e.g.* deca[CRU].

II-7.3.1.3 *Use of the term catena*

A linear (one-dimensional) polymer is indicated by the italicized prefix *catena* added to the name of the polymer formed according to Sections II-7.3.1.1 and II-7.3.1.2 (Note 7l), *e.g. catena*-poly[CRU]. The term *catena* is also used in Chapter II-5.

II-7.3.1.4 *Specification of end groups*

End groups of a polymer molecule may be specified, if desired, by appropriate prefixes identified by the Greek letters α and ω, which are added to the name of the polymer formed as given by Sections II-7.3.1.1 or II-7.3.1.2 and II-7.3.1.3, *e.g.* α (end group)-ω-(end group)--*catena*-poly[CRU]. For details see Section I-8.3.8 (Note 7d).

> Note 7j. Linear (chain), crosslinked, branched, *etc.*
> Note 7k. Although the rules for naming linear organic polymers (Notes 7a and 7b) do not provide for describing dimensional structure, such specification would allow the prefix poly to be a very general descriptor having no structural implications other than the presence of a number of constitutional repeating units.
> Note 7l. The prefix *catena* is consistent with the existing rules for naming coordination compounds with extended structures (Note 7d). In mineralogy and geochemistry, silicate chains have been denoted by the prefix *ino* and the prefixes *phyllo* and *tecto* are used for sheet (two-dimensional) and three-dimensional structures, respectively (Note 7d). The term *catena* should not be confused with terms such as catenane, or catena compounds, used to describe interlocking organic ring compounds.

II-7.3.2 Seniority rules for selection of a preferred CRU

Many regular single-strand and quasi-single-strand inorganic or coordination polymers can be represented as multiples of a repeating unit most conveniently represented as a series of smaller subunits. The following recommendations are concerned with various seniority considerations necessary for the derivation of a preferred CRU. Refinements to these basic recommendations and their application to specific polymers are illustrated in the sections which follow.

II-7.3.2.1 *Choice of the senior constituent subunit*

(i) The constituent subunit of highest seniority, *i.e.* the first subunit to be cited, in the preferred CRU of an inorganic or coordination polymer must contain one or more central atoms; bridging groups between central atoms in the backbone of the polymer cannot be senior subunits (see Note 7m).

(ii) When there are two (or more) central atoms in a CRU of a linear inorganic or coordination polymer, the senior subunit is that containing the central atom occurring latest in the general element sequence shown in Table II-1 (page 5) (Note 7n).

(iii) When a further choice is needed for the selection of a senior subunit in a CRU of a linear inorganic or coordination polymer, preference is given, in order, to:

(a) a polynuclear coordination centre, in order of decreasing number of central atoms, provided the use of polynuclear centres as a subunit of a CRU is necessary (see Section II-7.2.3);

(b) the central atom or coordination centre with the greatest number of attached donor atoms, excluding donor atoms of bridging ligands in the backbone of the polymer chain;

(c) the central atom or coordination centre the name of which, including ligands and any multiplying prefixes but not the bridging ligands in the backbone of the polymer chain, occurs earliest in the alphabet.

II-7.3.2.2 *Choice of the preferred direction along the polymer chain for the sequential citation of the constituent subunits of the CRU*

After the senior subunits of the CRU have been determined by satisfying all of the seniority considerations of Section II-7.3.2.1, the following general principles are applied in order, where applicable:

(i) When it is possible to have either a single-strand or a quasi-single-strand CRU, the preferred direction is that giving a single-strand CRU (Note 7o).

Note 7m. This is consistent with the principles of coordination nomenclature in which the emphasis is always on the coordination centre. There is always at least one coordination centre in each inorganic or coordination polymer.

Note 7n. Note that this seniority order is not the same as that for hetero atoms given in the organic polymer recommendations (Note 7b).

Note 7o. This principle is quite analogous to that of minimizing free valencies of constitutional repeating units in naming linear organic polymers when it is necessary to choose between a bivalent and a higher valent CRU after all factors concerned with the determination of subunit of highest seniority have been observed (Note 7b).

Example (M = central atom)

$$\left(\!M\!\genfrac{}{}{0pt}{}{\diagup S\diagdown}{\diagdown S\diagup}\!C\!-\!\underset{H_2}{C}\!-\!O\!\right)_{\!n} \quad \text{not} \quad \left(\!M\!-\!O\!-\!\underset{H_2}{C}\!-\!C\genfrac{}{}{0pt}{}{\diagup S\diagdown}{\diagdown S\diagup}\right)_{\!n}$$

(ii) *Shortest path.* The preferred direction along a polymer chain for the sequential citation of the constituent subunits of the CRU is the direction that leads first through the shortest path from the subunit of highest seniority to a subunit of equal seniority or to a subunit of next highest seniority. The length of the path between these subunits is the number of atoms in the most direct continuous chain of atoms from one unit to the other.

Examples (M = central atom)

1.
$$\left(\!M\!-\!\underset{}{\overset{R}{\underset{|}{S}}}\!-\!M'\!-\!\underset{H_2}{N}\!-\!\underset{H_2}{N}\!\right)_{\!n} \quad \text{is preferred to} \quad \left(\!M\!-\!\underset{H_2}{N}\!-\!\underset{H_2}{N}\!-\!M'\!-\!\overset{R}{\underset{|}{S}}\right)_{\!n}$$

The one-atom path through the thiolate ligand is preferred to the two-atom path through the hydrazine ligand.

2.
$$\left(\!M\!-\!\underset{\underset{\underset{S}{\parallel}}{\underset{C}{|}}}{N}\!-\!M'\!-\!N\!=\!C\!=\!S\!\right)_{\!n} \quad \text{is preferred to} \quad \left(\!M\!-\!N\!=\!C\!=\!S\!-\!M'\!-\!\underset{\underset{\underset{S}{\parallel}}{\underset{C}{|}}}{N}\right)_{\!n}$$

The one-atom path through the nitrogen atom of the thiocyanato ligand is preferred to the three-atom path through all of the atoms of the thiocyanato ligand.

3.
$$\left(\!M\!-\!O\!-\!\text{[quinoline]}\!-\!N\!\rightarrow\!M'\!\leftarrow\!NH_2\!-\!\text{[naphthalene]}\!-\!NH_2\!\right)_{\!n}$$

is preferred to

$$\left(\!M\!\leftarrow\!NH_2\!-\!\text{[naphthalene]}\!-\!NH_2\!\rightarrow\!M'\!\leftarrow\!N\!-\!\text{[quinoline]}\!-\!O\!\right)_{\!n}$$

The five-atom path through the 5-quinolinolato ligand is preferred to the six-atom path through the 1,4-diaminonaphthalene ligand.

(iii) When there are paths of equal (shortest) length between two subunits of equal highest seniority or between a subunit of highest seniority and a subunit ranking next in seniority, the preferred direction is determined by the kinds of structures and atoms included in the path and does not depend on actual names for subunits used in the final CRU name, unless there is no other choice remaining.

(a) For the selection of a preferred path, the same principles are used as in the organic polymer recommendations (Note 7b), in which the fundamental order of seniority is:

(1) heterocycles,
(2) acyclic heteroatoms,
(3) carbocycles, and
(4) acyclic carbon atoms or chains.

Seniority priority within these classes as given by the recommendations for organic nomenclature are also followed (Notes 7b and 7p).

Examples: (M = central atom)

4.

The heterocycle is preferred to the acyclic heterochain.

5.

The Cl ligand is preferred to the O ligating atom.

(b) Substituents on atoms or groups in the path are used to determine priority between otherwise identical paths according to the principles in Note 7b.

(c) If a further choice is necessary between otherwise identical paths, the preferred path leads by the shortest path, in the sense of Section II-7.3.2.2(ii), from the senior subunit to the most preferred structural feature in the path.

Examples: (M = central atom)

6. $-(-M-N\equiv C-)_n-$ not $-(-M-C\equiv N-)_n-$

The hetero atom N is preferred to the carbon atom, and hence is preferred for citation closest to the senior subunit, M, in the preferred CRU.

Note 7p. It is important to note that the seniority order for hetero atoms in ligands prescribed here is not the same as the seniority order for coordination centres [Section II-7.3.2.1(ii)].

7.

$$\left(-M-N\underset{\substack{\\}}{\overset{\substack{\\}}{\bigcirc}}-O-\right)_n \quad \text{not} \quad \left(-M-O-\underset{\substack{\\}}{\overset{\substack{\\}}{\bigcirc}}-N-\right)_n$$

The heterocycle is preferred to the hetero atom O, and hence is preferred for citation closest to the senior subunit, M, in the preferred CRU.

(d) If a further choice is needed, the preferred path contains the ligand whose name occurs earliest in the alphabet.

II-7.3.3   **Regular single-strand inorganic and coordination polymers**

Regular single-strand inorganic and coordination polymers are named by inserting the name of the preferred constitutional repeating unit into the appropriate general polymer name given in Section II-7.3.1.

II-7.3.3.1   *CRUs with homoatomic backbones*

CRUs with homoatomic backbones are named by citing each mononuclear central atom, together with its side groups, if any, named as ligands.

Examples: (Note 7q)
1. $(-S-)_n$
    catena-poly[sulfur] (Note 7d)

2. $\left(-\underset{\underset{CH_3}{|}}{\overset{\overset{CH_3}{|}}{Sn}}-\right)_n$

    catena-poly[dimethyltin]

3. $\left(-\underset{\underset{F}{|}}{\overset{\overset{F}{|}}{Si}}-\underset{\underset{CH_3}{|}}{\overset{\overset{CH_3}{|}}{Si}}-\right)_n$

    catena-poly[(difluorosilicon)(dimethylsilicon)] (the subunit with the alphabetically earliest side group is the senior subunit)

Note 7q. According to the recommendations for naming linear organic polymers (Notes 7a and 7b), these inorganic homoatomic polymers could be named: (1) poly(sulfanediyl); (2) poly(dimethylstannylene) or poly(dimethylstannanediyl); (3) poly(1,1-difluoro-2,2-dimethyldisilane-1,2-diyl).

POLYMERS

II-7.3.3.2 *Constitutional repeating units with backbones consisting of a central atom and one bridging ligand*

These are named by citing the central atom prefixed by its associated non-bridging ligands followed by the name of the bridging ligand prefixed by the Greek letter μ (Note 7r).

Examples:

1. $-\!\!-\!\!\left(\!\!-\!\!\text{N}\!=\!\!=\!\!\text{S}\!-\!\!\right)_{\!n}\!\!-\!\!-$

    catena-poly[nitrogen-μ-thio] {see II-7.3.3.2(i) below}

2. $\left(\!\!\begin{array}{c}\text{NH}_3\\|\\-\text{Zn}\!-\!\text{Cl}-\\|\\\text{Cl}\end{array}\!\!\right)_{\!n}$

    catena-poly[(amminechlorozinc)-μ-chloro]

3. $\left(\!\!\begin{array}{c}\text{S=C(NH}_2)_2\\|\\-\text{Ag}\!-\!\text{Cl}-\end{array}\!\!\right)_{\!n}$

    catena-poly[[(thiourea-*S*)silver]-μ-chloro]

    (i) If there is a choice for the central atom, the element occurring later in the general element sequence (Table II-1) is the central atom.

Examples:

4. $\left(\!\!\begin{array}{c}\text{Ph}\\|\\-\text{Si}\!-\!\text{O}-\\|\\\text{Ph}\end{array}\!\!\right)_{\!n}$

    catena-poly[(diphenylsilicon)-μ-oxo] (see Note 7s)

5. $\left(\!\!\begin{array}{c}\phantom{\text{B}}\text{CH}_3\\\phantom{\text{B}}|\\-\text{B}\!-\!\text{N}-\\\text{H}_2|\\\phantom{\text{B}}\text{CH}_3\end{array}\!\!\right)_{\!n}$

    catena-poly[(dihydridoboron)-μ-(dimethylamido)]

Note 7r. Note that the bridging ligand is not included with the other ligands attached to the central atom. However, bridging ligands of polynuclear coordination centres that are not backbone units of the polymer are added in their usual position in the name of the polynuclear coordination centres (see Section II-7.3.5).

Note 7s. According to the recommendations for linear organic polymers (Note 7a and 7b), this polymer would be oriented and named poly[oxy(diphenylsilylene)].

6.

$$\left(\begin{array}{c} \text{OEt} \\ | \\ \text{P}=\text{N} \\ | \\ \text{OEt} \end{array}\right)_n$$

catena-poly[(diethoxophosphorus)-μ-nitrido] (see Note 7t).

(ii) Italicized element symbols indicating the coordinating atoms of bridging ligands in the backbone are cited in the order of the direction of citation of the CRU and are separated by a colon. Hence element symbols cited before the colon refer to the central atom occurring just before the bridging ligand in the CRU and element symbols cited after the colon refer to the central atom occurring immediately after the bridging ligand in the CRU or in the polymer chain.

Examples:

7. $(-\text{SnMe}_3-\text{S}=\text{C}=\text{N}-)_n$
   catena-poly[(trimethyltin)-μ-(thiocyanato-*S*:*N*)]

8. $(-\text{Ag}-\text{NC}-)_n$
   catena-poly[silver-μ-(cyano-*N*:*C*)], not *catena*-poly[silver-μ-(cyano-*C*:*N*)] [see Section II-7.3.2.2.(iii)(c)]

9.

$$\left(\begin{array}{c} \text{Ph} \\ | \\ -\text{Sn} \\ | \\ \text{Ph} \end{array} \begin{array}{c} \text{O} \\ \diagdown \\ \diagup \\ \text{O} \end{array} \text{C}-\overset{\text{H}_2}{\text{C}}-\text{S}-\right)_n$$

catena-poly[(diphenyltin)-μ-[mercaptoacetato(2–)-*O*,*O*′:*S*]
(direction of citation chosen by Section II-7.3.2.2)

When a choice of direction for citation of the constituent subunits in the CRU remains after application of the principles in Section II-7.3.2.2, the direction is chosen so that the italicized element symbols denoting coordinating atoms of the bridging ligand are cited in lowest alphabetical order (see Example 6 in Section II-7.3.3.3).

II-7.3.3.3 *CRUs consisting of more than one central atom and no more than one bridging ligand between each central atom of the polymer*

These are named by extending the principles of Section II-7.3.3.2. The senior central atom is selected according to Section II-7.3.2.1, and the direction of citation is determined by Section II-7.3.2.2.

Note 7t. According to the recommendations for linear organic polymers (see Notes 7a and 7b), this inorganic polymer would be oriented and named poly[nitrilo(diethoxyphosphoranylidyne)].

Examples:

1.

$$\left(\begin{array}{c} \text{ONO}_2 \\ | \\ \text{-Hg-Hg-NH}_2 \quad \text{NH}_2- \\ | \\ \text{ONO}_2 \end{array} \begin{array}{c} \\ \diagup\!\!\!\!\bigcirc \\ \end{array} \right)_n$$

catena-poly[(nitratomercury)(nitratomercury)-μ-(*o*-phenylenediamine-*N*:*N*′)(*Hg*–*Hg*)]
(not *catena*-poly[(nitratomercury)-μ-(*o*-phenylenediamine-*N*:*N*′)(nitratomercury)-
-(*Hg*–*Hg*)]; Section II-7.3.2.2(ii)

2.

$$\left(\begin{array}{ccccccccc} \text{NH}_3 & & \text{S} & & \text{NH}_3 & & \text{S} & \text{S} & \\ | & & || & & | & & || & || & \\ \text{-Cu-O-C-O-Cu-O-C-C-O-} \\ | & & & & | & & & & \\ \text{NH}_3 & & & & \text{NH}_3 & & & & \end{array} \right)_n$$

catena-poly[(diamminecopper)-μ-[thiocarbonato(2–)-*O*:*O*′](diamminecopper)-μ-
-{dithiooxalato(2–)-*O*:*O*′}]
[direction selected according to Section II-7.3.2.2(ii)]

3.

$$\left(\begin{array}{ccc} \text{CN} & & \text{NH}_3 \\ | & & | \\ \text{-Ni-CN-Cu-NC-} \\ | & & | \\ \text{CN} & & \text{NH}_3 \end{array} \right)_n$$

catena-poly[{bis(cyano-*C*)nickel}-μ-(cyano-*C*:*N*)-(diamminecopper)-μ-(cyano-*N*:*C*)]
[senior subunit chosen according to Section II-7.3.2.1(ii)]

4.

$$\left(\begin{array}{cccc} \text{H}_2\text{N} & \text{NH}_2 & \text{SCN} & \\ \diagdown & \diagup & | & \\ \text{-Cu} & \text{-NCS-Hg-SCN-} \\ \diagup & \diagdown & | & \\ \text{H}_2\text{N} & \text{NH}_2 & \text{SCN} & \end{array} \right)_n$$

catena-poly[{bis(ethane-1,2-diamine-*N*,*N*′)copper}-μ-(thiocyanato-*N*:*S*)-
-[bis(thiocyanato-*S*)mercury]-μ-(thiocyanato-*S*:*N*)]

5.

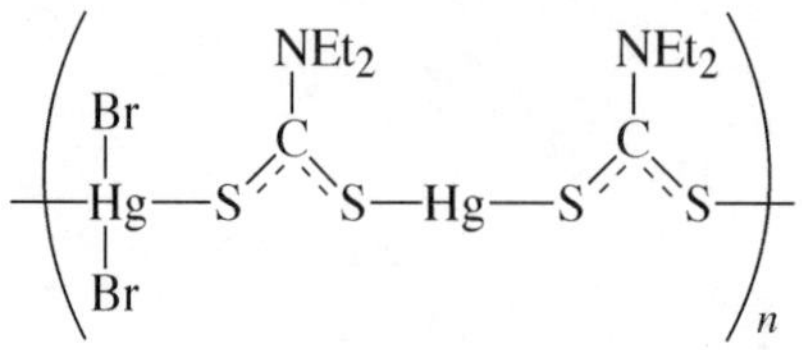

[structure: -(Hg(Br)₂-S-C(NEt₂)=S-Hg-S-C(NEt₂)=S-)_n]

catena-poly[(dibromomercury)-μ-(diethyldithiocarbamato-*S*:*S*′)-mercury-μ-(diethyl--dithiocarbamato-*S*:*S*′)]
[senior subunit chosen according to Section II-7.3.2.1(iii)(b)]

6.

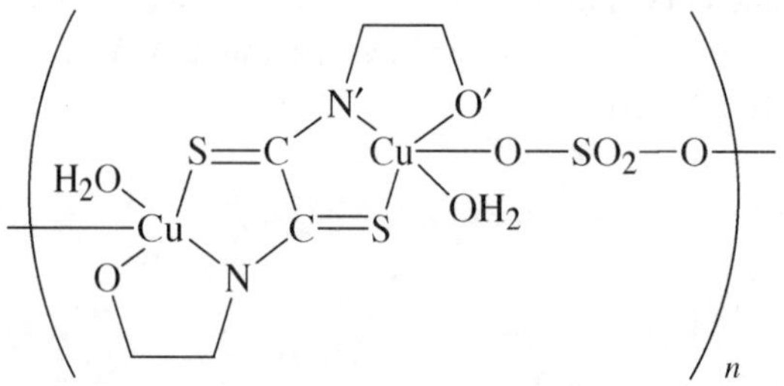

[structure of aquacopper dithiooxamido sulfato polymer]

catena-poly[(aquacopper)-μ-[*N*,*N*′-bis(2-hydroxyethyl)dithiooxamido(2–)--*N*,*O*,*S*′:*N*′*O*′*S*](aquacopper)-μ-[sulfato(2–)-*O*:*O*′]] (see Note 7u)

Multiple bridging ligands between the same pair of central atoms are cited in alphabetical order, each preceded by the Greek letter μ, and all are enclosed in square brackets to reduce the possibility of misinterpretation.

7.

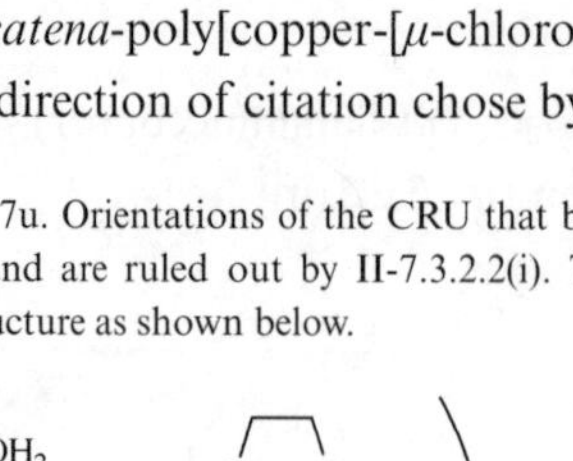

[structure: -(Cu-Cl-Cu-Cl-)_n with two Et-S-S-Et bridges]

catena-poly[copper-[μ-chloro-bis-μ-(diethyl disulfide-*S*,*S*′)]-copper-μ-chloro]
[direction of citation chose by Section II-7.3.2.2(i)] (Note 7v)

Note 7u. Orientations of the CRU that begin with either copper central atom and proceed first through the sulfato ligand are ruled out by II-7.3.2.2(i). These orientations produce the less-preferred quasi-single-strand polymer structure as shown below.

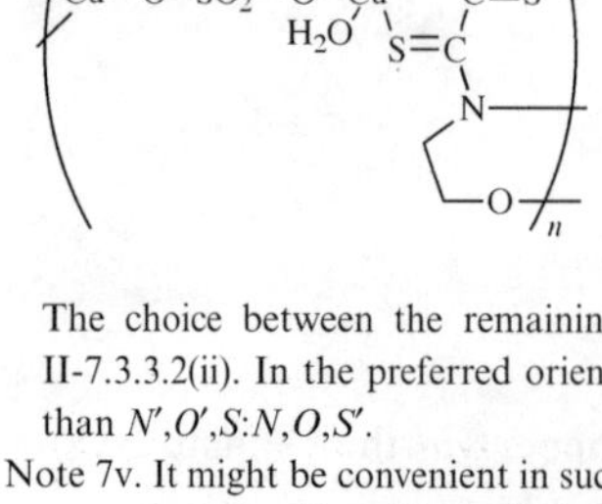

[structure showing alternative orientation]

The choice between the remaining two orientations is made according to the second paragraph of Section II-7.3.3.2(ii). In the preferred orientation, the letter locant order for the bridging ligand, *N*,*O*,*S*′:*N*′,*O*′,*S*, is lower than *N*′,*O*′,*S*:*N*,*O*,*S*′.

Note 7v. It might be convenient in such cases to treat the two copper central atoms as a binuclear complex as follows: *catena*-poly[[μ-chloro-bis-μ-(diethyl disulfide-*S*:*S*′)-dicopper]-μ-chloro].

POLYMERS

8.

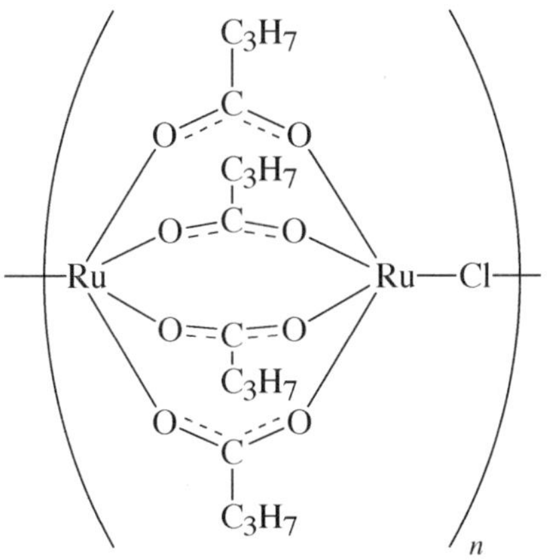

catena-poly[[ruthenium-tetrakis-μ-(butanoato-*O,O'*)-ruthenium(*Ru—Ru*)]-μ-chloro] [direction of citation chosen by Section II-7.3.2.2(i); (Note 7w)]

II-7.3.4 **Regular quasi-single-strand coordination polymers**

Regular quasi-single-strand coordination polymers are named by inserting the name of the preferred constitution repeating unit into the appropriate general polymer name as given by Section II-7.3.1.

II-7.3.4.1 *CRUs with backbones consisting of one central atom and two or more bridging ligands or chelating ligands*

CRUs with backbones consisting of one central atom and two or more bridging ligands, alike or different, or chelating ligands, are named by citing the name of the central atom, prefixed by the names of its associated non-bridging ligand(s), followed by the names of the bridging ligands each prefixed by the Greek letter μ. The number of identical bridging ligands, if more than one, is indicated by an appropriate numerical prefix; different bridging ligands are cited in alphabetical order and all are enclosed in appropriate enclosing brackets.

Examples:

1.

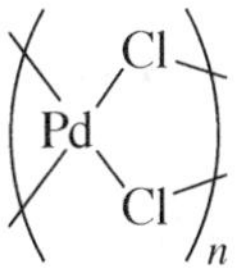

catena-poly[palladium-di-μ-chloro]

Note 7w. It is also possible to treat the two ruthenium atoms as a binuclear centre, giving the name *catena*-poly[[tetrakis-μ-(butanoato-*O:O'*)-diruthenium(*Ru—Ru*)]-μ-chloro].

2.

catena-poly[titanium-tri-μ-chloro]

3.

catena-poly[silicon-di-μ-thio]

4.

catena-poly[beryllium-bis-μ-[diphenylphosphinato(1–)-O,O']]

5.

catena-poly[platinum(μ-bromo-μ-chloro)]

6.

catena-poly[zinc-μ-[2,5-dihydroxy-p-benzoquinonato(2–)-$O^1,O^2:O^4,O^5$]]

7.

catena-poly[beryllium-μ-[1,14-diphenyltetradecane-1,3,12,14-tetronato(2–)-
-$O^1,O^3:O^{12},O^{14}$]]

8.

catena-poly[nickel-μ-[dithiooximidato(2–)-$N^1,S^2:N^2,S^1$]]

9.

catena-poly[cadmium-μ-[[5,5′-[biphenyl-4,4′-diylbis(methylidynenitrilo)]di-8-
-quinolinolato](2–)-$N^1,O:N^{1'},O'$]

II-7.3.4.2   *CRUs containing more than one central atom*

CRUs containing more than one central atom are named by extending the principles of Section II-7.3.4.1. The senior central atom is selected according to Section I-7.3.2.1, and the direction of the citation is determined by Section II-7.3.2.2.

Examples:

1.

catena-poly[(oxovanadium)-di-μ-hydroxo-(oxovanadium)-[μ-(quinolin-8-olato-$N:O$)-μ-
-(quinolin-8-olato-$O:N$)]

[the direction of citation is determined by Section II-7.2.3.2.2(ii)]

2.

catena-poly[copper-di-μ-chloro-copper-μ-[2,2'-(azo-N:$'$)dipyridine-N:N']]
[the direction of citation is determined by Section II-7.3.2.2(ii)]

II-7.3.5 Single-strand and quasi-single-strand coordination polymers with polynuclear coordination centres

Single-strand and quasi-single-strand coordination polymers with one polynuclear coordination centre are named in much the same manner as coordination polymers having only one mononuclear centre. The polynuclear centre is the senior subunit and begins the citation of the subunits in the name of the CRU. Both positions on the polynuclear complex where the bridging ligand is attached are indicated by numerical locants, inserted between the name of the polynuclear centre and the names of the bridging ligand, and separated by a colon; locants before the colon refer to the preceding polynuclear centre in the CRU and the locants following the colon refer to the next polynuclear centre in the polymer chain.

Examples:
1. {–[-*octahedro*-W$_6$(μ-Br$_8$)(2,3,4,5-Br$_4$)–]–(6:1)–(Br$_4$)–}$_n$
catena-poly[octa-μ-bromo-2,3,4,5-tetrabromo-*octahedro*-hexatungsten)-6:1-μ--[tetrabromido(2–)]]

2.

catena-poly[[1,2:1,3:2,3-tri-μ-iodo-1,2,3,3-tetraiodo-*triangulo*-trirhenium(3Re—Re)]--2,2:1,1-di-μ-iodo] (Note 7x)

II-7.3.6 Regular single-strand and quasi-single-strand inorganic and coordination polymers with ionic CRUs

Regular single-strand and quasi-single-strand inorganic and coordination polymers with ionic CRUs are named in the same general manner as described in Sections II-7.3.3, II-7.3.4 and II-7.3.5. The charge of the CRU may be indicated by a charge number cited after the name of

---

Note 7x. Numbering rules for polynuclear complexes have not been fully defined. The numbering shown here is arbitrary and only for convenience in defining the structure of this polymer in this Chapter.

the ionic portion of the CRU. Oxidation numbers may be used to denote the oxidation state of the central atom; if so, they are attached to the name of the central atom in the usual manner.

Examples (see Note 7y):

1.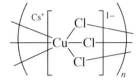

catena-poly[caesium[cuprate-tri-μ-chloro](1−)] or
catena-poly[caesium[cuprate(II)-tri-μ-chloro]]

2.

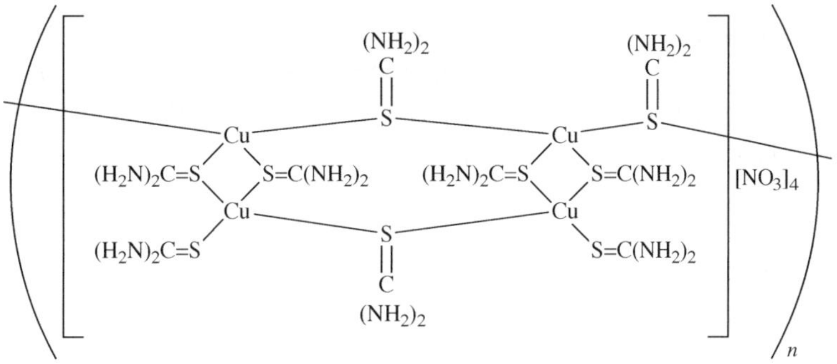

catena-poly[[bis-μ-(thiourea-*S,S*)-2-(thiourea-*S*)dicopper]-1,2:1,2-[bis-μ-(thiourea--*S,S*)][bis-μ-(thiourea-*S,S*)-2-(thiourea-*S*)dicopper]-1:1-μ-(thiourea-*S:S*)](4 +)tetranitrate]
or
catena-poly[[bis-μ-(thiourea-*S,S*)-2-(thiourea-*S*)dicopper(I)]-1,2:1,2-[bis-μ-(thiourea--*S,S*)][bis-μ-(thiourea-*S,S*)-2-(thiourea-*S*)dicopper(I)]-1:1-μ-(thiourea-*S:S*)] tetranitrate]

II-7.3.7   **Stereochemical configuration of a CRU**

The stereochemical configuration of a CRU consisting of a mononuclear central atom and one bridging ligand may be designated by suitable prefixes cited before the appropriate complete polymer name (Note 7z).

Note 7y. From an inorganic viewpoint, it might be better to consider these polymers as salts of polymeric ions as illustrated below:

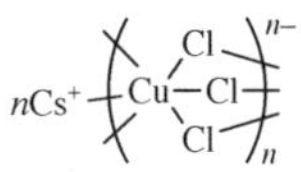

Caesium *catena*-poly[cuprate-tri-μ-chloro](1−) or caesium *catena*-poly[cuprate(II)-tri-μ-chloro].

Note 7z. An alternative format in which the stereochemical prefix is inserted between the prefix poly and the name of the CRU for both examples under this recommendation would be more consistent with inorganic nomenclature practice, as shown by the names: (i) *catena*-poly[*cis*-[(difluorogold)-μ-fluoro]]; (ii) *catena*-poly[*trans*-[dipotassium [(tetrafluoroaluminate)-μ-fluoro](2−)]] or *catena*-poly[*trans*[dipotassium [tetrafluoroaluminate(III)]-μ-fluoro]]. However, the recommendation in Section II-7.3.4 is in accord with the stereochemical notation recommendation for organic polymers [*Pure Appl. Chem.*, **53**, 733 (1981)].

Examples:

1.

cis-*catena*-poly[(difluorogold)-μ-fluoro]

2.

trans-*catena*-poly[dipotassium[[tetrafluoroaluminate)-μ-fluoro](2–)]] or
trans-*catena*-poly[dipotassium[[tetrafluoroaluminate(III)]-μ-fluoro]]
(inorganic nomenclature practice would not require the use of the oxidation number for aluminium)

II-7.3.8 End groups of linear inorganic or coordination polymers

End groups of linear inorganic or coordination polymers are specified by prefixes cited in front of the name of the polymer (see Section II-7.3.1.4).

II-7.3.8.1 *Use of the terms α and ω*

The groups attached to the first constituent subunit of the preferred CRU, *i.e.* the senior coordinaton centre written as the left terminal subunit in the CRU, are named as ligands and designated by the Greek letter α.

End groups attached to the other terminal subunit of the preferred CRU are named as ligands if attached to a central atom or, if attached to a bridging ligand, are named as a central atom by the usual principle of coordination nomenclature and are designated by the Greek letter ω.

Examples:

1. Cl–(–S–)$_n$–H
   α-chloro-ω-hydrido-*catena*-poly[sulfur]

2.

α,α-diaqua-ω-[[2,5-dihydroxy-*p*-benzoquinonato(1–)-O^1,O^2]zinc]-*catena*-poly[zinc-μ-[2,5--dihydroxy-*p*-benzoquinonato(2–)-O^1,O^2:O^4,O^5]]

II-7.3.8.2 *Choice of end groups*

When a choice is necessary to determine ligands to be cited as an end group or to be included in the CRU, the ligand chosen as the α end group is that whose name occurs earliest in the alphabet.

Example:

1.

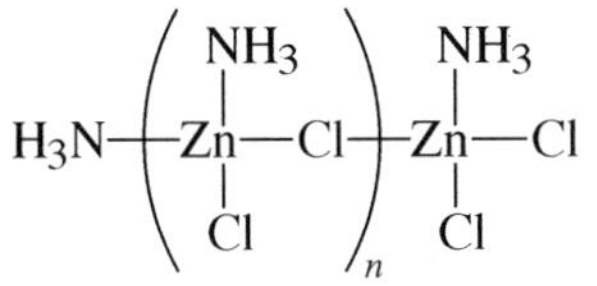

$$H_3N-\left(Zn(NH_3)(Cl)-Cl\right)_n-Zn(NH_3)(Cl)-Cl$$

α-ammine-ω-(amminedichlorozinc)-*catena*-poly[(amminechlorozinc)-μ-chloro]

II-7.3.8.3 *Ionic end groups*

End groups that may be considered ionic are named by the principles of coordination nomenclature. The charge is indicated by a charge number cited at the end of the complete polymer name.

Examples:

1. $[O-(-MoO_2-O-)_n-MoO_3]^{2-}$
   [α-oxo-ω-(trioxomolybdate)-*catena*-poly[(dioxomolybdenum)-μ-oxo]](2–)

2.

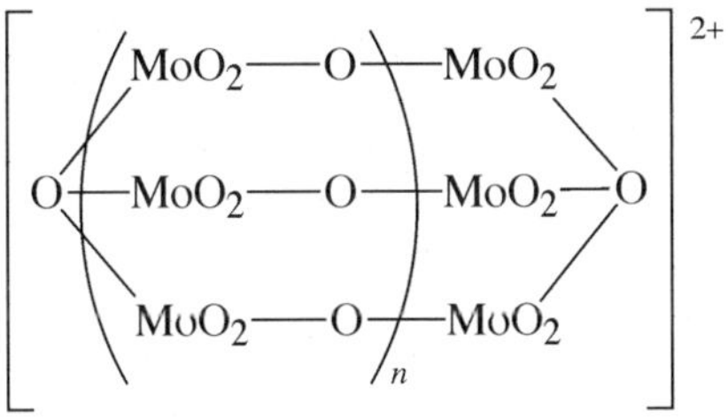

$$\left[\begin{array}{c}\text{three chains of } (MoO_2-O)_n \text{ linked by terminal O atoms}\end{array}\right]^{2+}$$

[α-μ_3-oxo-ω-[μ_3-oxo-tris(dioxomolybdenum)]tris[*catena*-poly-[(dioxomolybdenum)-μ--oxo]](2+) (Note7aa)

Note 7aa. This is a regular single-strand polymer consisting of three chains linked by a terminal oxo end group.

Subject Index

additive nomenclature
   chains and rings 63–94
   nitrogen hydrides 54
alpha (α) symbol
   for polymer end groups 110, 124–125
   and stereochemistry of tetrapyrrole
      complexes 39, 45
aluminium, in polyanions 13, 15
Anderson-type polyanions 11
arsenic, in polyanions 11, 19–20
assemblies, of chains and rings 79–80, 83–84, 88–89
azane (also ammonia) 55
   names of derivatives 55–59
      anionic ligands 59
      anions 56
      cations 55
      ligands 57–58

bacteriochlorins 48, 52
beta (β) symbol, and stereochemistry of
   tetrapyrrole complexes 39, 45
bilanes 50–51, 53
bilirubins 50
   complexes 43

cage compounds 89–91
catena, as affix
   for chain compounds 63–71, 86, 88, 92
   for single strand polymers 107, 110, 114–125
catenate ions 63, 70–71
   as ligands 71
catenium ions 63, 69
chains and rings 62–94
   assemblies,
      descriptors 83–84
      name construction 88–89
      numbering 79–80
   branched chains,
      graphs 75–78, 81
      name construction 86
      nodal descriptors 76
      nodal nomenclature 75
   indication of charge 70
   monocycles 71–75
      anions 74–75
      atom numbering 71–74
      cations 74
      ligand names 85–86

polycycles 86–88
   cages, nodal nomenclature 89–91
   clusters 94
   coordination compounds 93
   graphs 76, 78, 82
   ions 91–92
   ligand names 92
   name construction 66–69
   position of isotopic modification 34
unbranched chains 63–71
   anions 70–71
   atom numbering 65–66
   cations 69
   choice of principal axis 64–65
   name construction 66–69
charge transfer,
   in graphite derivatives 96, 98
   in homopolyanions 7
chlorins 48, 52
   complexes 38
chlorophylls 49–50
   complexes 42
CIP (Cahn, Ingold and Prelog) rules,
   tetrapyrrole complexes 37, 39–41
clusters,
   position of isotopic modification 34
   ring nomenclature 94
configuration index,
   nitrogen hydride complexes 57–58
   tetrapyrrole complexes 57–58
constitutional repeating units (CRUs),
   backbones,
      homoatomic 114
      more than one central atom 121–122
      one central atom and one bridge 115–116
      one central atom and two or more bridges 119–121
      two or more central atoms and one bridge 116–119
constituent subunits 107
in graphite intercalation compounds 97, 101–102
identification 107–108
ionic units 122–123
naming 109–110
   bridging ligands 109
orientation 108–109
polynuclear coordination centres 122
stereochemical configuration 123–124

SUBJECT INDEX

coordination complexes, chain and ring nomenclature 93–94
coordination nomenclature
   and inorganic polymers 106, 111
   tetrapyrrole complexes 38–39
corrins 52
corroles 52
cyclate ions 63, 74, 90–91, 93
   as ligands 74–75, 91
cyclium ions 63, 74, 91–92

Dawson structure for polyanions 20
   isomers 21
   and Keggin structure 20
def, descriptor for isotopic modification 33
defect structures,
   graphite intercalation compounds 96
   polyanions 18–20
descriptors,
   for chains and rings 62–94
     atoms 63, 71, 79
     graphs 75–84
     nodal descriptors 76–94
   for isotopically modified compounds 32–33
     general labelling 32
     isotopically deficient compounds 33
     uniform labelling 32
   for stereochemistry of tetrapyrrole complexes 39–41
diazane (also hydrazine) 55
   names of derivatives 55–61
     anionic ligands 59–60
     anions 56
     cations 55
     ligands 57–58
     organic ligands 61
diazene (also diimide) 55
   names of derivatives 55–61
     anionic ligands 60
     anions 56
     cations 55
     ligands 57–58
     organic ligands 61
donor atom symbols,
   and nitrogen hydride ligands 57–58
   and tetrapyrrole complexes 38

electrical conductivity, and graphite intercalation 96
element seniority sequence 5
   chains and rings 66, 79
element substituent group names, chains and rings 63–64, 71
end groups, in polymers 110, 124–125
epitaxy, graphite intercalation compounds 101–102

formulae,
   graphite intercalation compounds 100–101
   isotopically modified compounds 26, 28, 30–31

Gard system, and graphite intercalation compounds 101
gen, descriptor for isotopic modification 32
graphs, as structural descriptors for rings and chains 75–84
graphene, as carbon layer in intercalation compounds 97–98, 100–101
graphite intercalation compounds 95–103
   carbon layers 96–97
   classification 98–99
     acceptor compounds 98
     donor compounds 98
     graphite salts 98
     heterostructures 99
     multi-intercalation 99
     number of components 98–99
   commensurate structures 101
   defects 96, 101
   electrical conductivity 96
   formulation 100–101
     bi-intercalation 100–101
     with solvated ions 10
   non-formula based description 99–10
   structural notation 101–103
     crystal lattice 101–102
     distance symbols 102–103
     and epitaxial layer 102
     stacking sequence 101–102
   ternary compounds 98, 100

heme (haem) complexes 42
hydride nomenclature, and nitrogen hydrides 54–55
hydrogen, names of atoms, ions and groups 25

intercalation compounds,
   graphite 95–103
   tantalum disulfide 97
iron, tetrapyrrole complexes 37, 39, 41–43, 45
isomerism, in polyanions 12–17, 21
isotopic modification 23–35
   classification 24–25
   isotopically deficient compounds 33, 35
   isotopically substituted compounds 26–27, 35
     definition 26
     formulae 26
     names 26–27
   locants 34
   non-selectively labelled compounds 33, 35
     definition 33
     non-molecular materials 33

SUBJECT INDEX

numbering 34–35
selectively labelled compounds 30–33, 35
   definition 30
   formulae 30–31
   general labelling 32
   mixed labelling 30
   multiple labelling 30
   names 31
   number of nuclides 31–32
   uniform labelling 32–33
specifically labelled compounds 28–30, 35
   definition 28
   formulae 28
   mixed labelling 30
   multiple labelling 29
   names 28–29
   single labelling 29

kappa (κ) convention,
   and nitrogen hydride ligands 57–58
   and tetrapyrrole complexes 38
Keggin structure, for polyanions 11–20
   and Dawson structure 20
   defect structures 18–20
   isomers 12–14
   ligand substitution 17
   and Lindqvist structure 15
   locant designators 12, 14
   reduction 18
   substituted compounds 16–17

ligands,
   bound to chains 85–86, 91
   from cages 92–93
   from chains 71
   from monocyclic rings 74–75
   order in tetrapyrrole complexes 38
Lindqvist structure 3, 6, 9–10, 15
locant designators,
   for chains and rings 63, 65, 69, 74, 81, 87
   Dawson structure 21
   for isotopically modified compounds 31, 34–35
   Keggin structure 12, 14
   Lindqvist structure 3

molybdenum,
   heteropolyanions 9–11
   homopolyanions 7
   Keggin polyanions 12, 16–19
mu (μ), symbol for bridging in polymers 109, 115, 118–119

names,
   branched chain and polycyclic compounds 84–94
      acyclics 86

      assemblies 88–89
      cages 89–91
      coordination compounds 93–94
      ions 91–92
      ligands 92–93
      polycycles 86–88
   chain compounds 66–69
      anions 70–71
      cations 69
   hydrogen atoms, ions and groups 25
   isotopically modified compounds 26–29, 31
   monocyclic compounds 71–75
      anions 74–75
      cations 74
   nitrogen hydrides 55
      anions 56
      cations 55
   nitrogen hydride ligands 57–58
      anions 58–60
      cations 60
      organic derivatives 61
   single strand polymers 105–125
      constitutional repeating units 109–10
   tetrapyrrole complexes 42, 46–53
niobium,
   heteropolyanions 7–10, 16
   homopolyanions 6–7
nitrogen hydrides and derivatives,
   anions,
      by hydron loss 56
      as ligands 58–60
   azane derivatives 55–61
   cations,
      by hydron addition 55
      indication of charge 55
   diazane derivatives 55–61
   diazene derivatives 55–61
   organic derivatives of ligands 61
   parent hydrides 54–5
   as ligands 57–58
nodal nomenclature,
   branched chains and polycycles 75–94
      acyclics 76, 81
      assemblies 79–80, 83–84, 88–89
      cages 89–91
      coordination compounds 93–94
      cycles 78, 82–83
      graphs 75–84
      ions 91–93
      name construction 84–94
      nodal descriptors 76–84, 86, 89–90
      nodal numbering 76–80
   organic compounds 75
nuclide symbols 25
   italicized locants 35
   order in formulae 25, 29–30
numbering systems,

SUBJECT INDEX

atoms in tetrapyrroles 37
for chains 65–66
for isotopically modified compounds 34–35
of nodes in branched chains and polycycles 76–80
for polyanions 2–4, 6
ring atoms 71
   direction 72–73
   procedure 73–74

omega (ω), symbol for polymer end groups 110, 124–125
oxidation state
   and chain and ring compounds 68
   and constitutional repeating units 123

phosphorus, in polyanions 19–22
phthalocyanins 52
   complexes 40–43
poly, as prefix in polymer nomenclature 110
polyanions 1–22
   with Anderson structure 10–11
   condensed structures 3
   Dawson structure 20–22
   designation of vertices 2
   with eighteen central atoms 20–22
   heterocentre polyanions 7–8, 16–17
   Keggin structure 11–20
   ligand substitution 8–9, 17
   mixed valence compounds 7
   numbering system 2, 4
   octahedron vertex designation 6
   preferred terminal skeletal plane 4
   reduction
      of heteropolyanions 8
      of homopolyanions 7
      of Keggin structure 18
   reference axes 3
   reference symmetry plane 4
   with seven central atoms 10–11
   with six central atoms 6–10
      heteropolyanions 7
      homopolyanions 6–7
   with twelve central atoms 11–20
polyhedral symbols,
   nitrogen hydride complexes 57–58
   tetrapyrrole complexes 39
porphyrinogen 52
porphyrins,
   complexes 36–9, 41–45
   numbering scheme 37
   trivial names 46–47, 52
priorities (also seniorities)
   constitutional repeating units 107–114
      choice of subunit 109–111
      direction along polymer chain 108–109, 111–112, 116

      naming 109–110
      organic units 113–114
      orientation 108
      shortest path 109, 112
   elements 5
   isotopically modified compounds 34
   in naming polyanions 3–6
   in nitrogen hydride ligand names 57
   nodal nomenclature for chains and rings,
      atom numbering 71–72
      element sequence 79, 89
      modules in assemblies 79–80
      node numbering 76
   tetrapyrrole complexes 39–41
pyrrins 53

reduction, of polyanions 7–8, 18

sapphyrin 52
seniorities (*see* priorities)
silicon, in polyanions 12–20
single strand polymers,
   end groups 124–125
      choice 125
      ionic 125
      use of α and ω, 124
   names 110
      numerical prefixes 110
      specification of end groups 110
      use of poly prefix 110
      use of square brackets 110
      use of term catena 110
   and organic nomenclature 105, 107, 110, 113–116
   quasi 119–123
      ionic constitutional repeating units 122–123
      more than one central atom 121–122
      one central atom and two or more bridges 119–121
      polynuclear coordination centres 122
   regular 114–119, 122–123
      homoatomic backbones 114
      ionic constitutional repeating units 122–123
      more than one central atom and one bridge 116–119
      one central atom and one bridge 115–116
      polynuclear coordination centres 122
square brackets,
   and isotopic modification 29, 31, 35
   and polymer nomenclature 110
stacking sequence, in graphite intercalation compounds 101–103
stereochemistry,
   configuration of constitutional repeating units 123–124

SUBJECT INDEX

descriptors for tetrapyrrole complexes 39–41, 45
substitutive nomenclature, and nitrogen hydrides 54

tetrapyrrole complexes 36–53
  atom numbering 37
  axial ligand designation 39
  coordination nomenclature 38–39
  less common structural types 43–45
  stereochemical descriptors 39–41

systematic names 37–41
trivial names 42, 46–53
tungsten,
  heteropolyanions 7–9
  homopolyanions 7, 10
  Keggin polyanions 11–20
    isomers 13–17

unf, descriptor for isotopic modification 32–33

vanadium, heteropolyanions 8, 16–17

RETURN TO:	CHEMISTRY LIBRARY		
100 Hildebrand Hall • 510-642-3753			
LOAN PERIOD 1	2		3
4	5 1-MONTH USE		

ALL BOOKS MAY BE RECALLED AFTER 7 DAYS.
Renewals may be requested by phone or, using GLADIS,
type **inv** followed by your patron ID number.

DUE AS STAMPED BELOW.

~~NON-CIRCULATING UNTIL:~~

AUG 3 1 2006

FORM NO. DD 10
2M 5-01

UNIVERSITY OF CALIFORNIA, BERKELEY
Berkeley, California 94720–6000